THE MILITARY COVENANT

The Military Covenant
Its Impact on Civil–Military Relations in Britain

SARAH INGHAM

Routledge
Taylor & Francis Group
LONDON AND NEW YORK

First published 2014 by Ashgate Publishing

2 Park Square, Milton Park, Abingdon, Oxon OX14 4RN
711 Third Avenue, New York, NY 10017, USA

Routledge is an imprint of the Taylor & Francis Group, an informa business

First issued in paperback 2016

British Library Cataloguing in Publication Data
A catalogue record for this book is available from the British Library

The Library of Congress has cataloged the printed edition as follows:
Ingham, Sarah.
The military covenant : its impact on civil-military relations in Britain / by Sarah Ingham.
pages cm
Includes bibliographical references and index.
ISBN 978-1-4724-2854-7 (hardback)
1. Civil-military relations—Great Britain. 2. Great Britain—
Armed Forces—Public relations. I. Title.
JN297.C58.I64 2014
322'.50941—dc23

2013050272

ISBN 978-1-4724-2854-7 (hbk)
ISBN 978-1-138-27230-9 (pbk)

Contents

List of Abbreviations

ADP	Army Doctrine Publication
AFOPS	Armed Forces Overarching Personnel Strategy
ARRSE	Army Rumour Service
BMD	*British Military Doctrine*
CDS	Chief of the Defence Staff
CGS	Chief of the General Staff
COBSEO	Confederation of Service Charities
CSR	Corporate Social Responsibility
DCDC	Development, Concepts and Doctrine Centre
ERG	External Reference Group
HCSC	Higher Command and Staff Course
MoD	Ministry of Defence
PJHQ	Permanent Joint Headquarters
RUSI	Royal United Services Institute
SDR	Strategic Defence Review
SDSR	Strategic Defence and Security Review
UAV	Unmanned Aerial Vehicle
UOR	Urgent Operational Requirement

Introduction

> Soldiers will be called upon to make personal sacrifices – including the ultimate sacrifice – in the service of the Nation. In putting the needs of the Nation and the Army before their own, they forego some of the rights enjoyed by those outside the Armed Forces. In return, British soldiers must always be able to expect fair treatment, to be valued and respected as individuals, and that they (and their families) will be sustained and rewarded by commensurate terms and conditions of service.
>
> In the same way the unique nature of military land operations means that the Army differs from all other institutions, and must be sustained and provided for accordingly by the Nation. This mutual obligation forms the Military Covenant between the Nation, the Army and each individual soldier; an unbreakable common bond of identity, loyalty and responsibility which has sustained the Army throughout its history.
>
> It has perhaps its greatest manifestation in the annual commemoration of Armistice Day, when the Nation keeps covenant with those who have made the ultimate sacrifice, giving their lives in action.
>
> Soldiering: The Military Covenant, Paragraph 103[1]

A covenant was a familiar concept to Cromwell's Ironsides in the New Model Army. *The Soldier's Catechism* of 1644 referred to the Solemn League and Covenant entered into by Parliament and the Scottish Presbyterians a year earlier: 'All Parliament's affairs have prospered well since the Covenant was taken, look at our victories and successes'.[2] For video-gamers after 2001, Covenant was an alien alliance, the chief antagonist of the *Halo* series.[3] However, from late 2006, a different covenant began to establish itself in the collective British consciousness. Today, the Military Covenant has become the prism through which all welfare or 'people' issues relating to Britain's Armed Forces community – serving personnel, their families and veterans – are seen and the context in which those issues are debated. More contentiously, it became the trope to convey the level of moral and material support that civilian society, both government and the public, gave

1 Army Doctrine Publications. Ministry of Defence. © Crown copyright 2014.

2 *The Soldier's Catechism*, 1644.

3 Of 20 British undergraduates contacted in 2011, 18 associated the term 'Covenant' with Halo.

to personnel on the frontline in Iraq and Afghanistan. Forster describes it as 'the cornerstone of Britain's civil–military compact'.[4]

Although the Military Covenant is now established in public policy, uncertainty lingers about its exact nature and provenance. Counter-intuitively, the Military Covenant is not an antique compact signed by the New Model's Commander-in-Chief and neither was it drawn up in Wellington's era, as the *Daily Mail* reported.[5] Like many traditions to be found in Britain, particularly those surrounding national institutions such as the monarchy and the Army, the Military Covenant was invented.[6] It was codified in a paragraph in *Soldiering: The Military Covenant* that was published in 2000, as part of the 'Moral Component' of British military doctrine. At its simplest, the Covenant is an understanding: in exchange for their service and sacrifice, soldiers will be supported by the nation and by their chain of command. Military doctrine is a living document, subject to regular review. In 2005, *Soldiering* was subsumed into the Army's capstone publication, *Land Operations*, and the paragraph that encapsulated the Military Covenant was revised.[7]

Without the operations in Iraq and Afghanistan, it is unlikely that the Military Covenant would have remained anything but an aspiration set down in Army doctrine, unknown to most soldiers, let alone to anyone outside the military. However, this was to change when the Covenant migrated to the civilian sphere.[8] The process began in late 2006 when the newly-appointed Chief of the General Staff, General Sir Richard Dannatt (now Lord Dannatt), gave a candid interview – or a carefully candid interview – in which he spoke about the Covenant and implied that the nation could be letting soldiers down.[9] In the following months, other senior officers framed welfare or people issues in the Covenant context.[10] The concept gained further traction in civilian society when the Prime Minister, Tony Blair, acknowledged it in a speech in January 2007.[11] In September, the Royal British Legion launched its *Honour the Covenant* campaign, aimed specifically at policy-makers, who were lobbied by the Legion's members. The accompanying media coverage raised further awareness of the concept.

The progress of the Military Covenant from the relative obscurity of an Army manual to being given formal recognition in statute as part of the 2011 Armed Forces Act is exceptional. In the light of its growing national recognition, a Parliamentary

4 Forster, 2006, pp. 1043–1057.

5 Our Boys Betrayed, *Daily Mail*, 14 September 2007.

6 Hobsbawm and Ranger, 1989.

7 Chapter 3.

8 The copyright on 'migration' in this context is Christopher Dandeker's.

9 Sarah Sands, *Daily Mail*, 12 October 2006.

10 E.g. Lt General Sir Frederick Viggers in Evans, *The Times*, 5 January 2007, p. 31; General Sir Michael Rose, American Forces Got their Money: Ours Have Been Betrayed, *The Independent on Sunday*, 31 December 2006, p. 28.

11 Blair, 12 January 2007.

question about whether there was Military Covenant appeared paradoxical.[12] Although Lord Attlee might have been obliquely attacking the Government for not fulfilling its obligations towards soldiers, his question could also have concerned the Covenant's validity and legitimacy. After all, it was not a legally enforceable contract, or to be found in the 2006 Armed Forces Act or Queen's Regulations, but an understanding based on the moral principle of reciprocity.

During the process of migration from doctrine to its entrenchment in the civilian sphere, the concept metamorphosed. The Covenant came to encompass first, all three Services and then, families and veterans. While the RAF and the Navy had followed the Army's example and both developed their own Service doctrine in the 1990s, neither had given the same level of attention to the 'Moral Component'. With some politicians and academics calling for the introduction of a tri-Service Covenant from late 2007, the extension of the concept's scope was a civilian rather than a Service initiative, reflecting the civilianisation of the Military Covenant and the Army's loss of ownership and control of the concept. Despite the statutory recognition given to the Armed Forces Covenant, civilian society continues to refer to the Military Covenant.

Emanating from military doctrine and promoted during two major combat operations, the Covenant remains chiefly associated with soldiers. It conceptualises a tripartite relationship between nation, Army and individual soldier, bound by an 'unbreakable common bond of identity, loyalty and responsibility', according to its original formalisation in Army doctrine in 2000. Within the Covenantal trinity is a lattice of bilateral relationships, with the 'nation' comprising both policy-makers and public. However, following migration, the Military Covenant came to be perceived by the civilian sphere as a bilateral agreement between the government and soldiers imbued erroneously with legal status, rather than a trilateral understanding based on reciprocal expectations and obligations. In the context of the relationship between government and military, the post-migration Covenant was described in relentlessly negative terms; 'well and truly broken', 'clearly out of kilter', 'almost beyond repair'.[13] With more than a decade of war underway on what some analysts considered was a peacetime defence budget, including almost three years of concurrent campaigns in Iraq and Afghanistan, the Blair-Brown Governments were perceived as failing to fulfil their side of the Covenant bargain. While soldiers were indeed making personal sacrifices, including 'the ultimate sacrifice', the belief became rooted that they and their families were not being 'sustained and rewarded by commensurate terms and conditions of service', as the Covenant demands.

An 'implicit contract based purely on expectations',[14] the Military Covenant became a shorthand term for all the shortcomings faced by an under-resourced and

---

12 Attlee, HL 21 April 2008 Col WA217.

13 David Cameron, 4 March 2008, Dannatt (Rayment, *Sunday Telegraph* 18 November 2007, p. 1), Edmonds and Forster, 2007, p. 13.

14 Campbell, 2008, p. 12.

overstretched military, encapsulating how the government was failing the Forces' community in general but frontline personnel in particular. Post-migration, it was initially linked to welfare issues, such as healthcare for the wounded, but it was subsequently associated with the frontline. The use of Snatch Land Rovers, a perceived helicopter shortage and the 2006 crash of Nimrod XV230 were all described in the Covenant context. As Tipping argues, it became 'a convenient portmanteau for a complex range of issues and relationships'.[15] In addition, the Covenant concerned not only the material support that the Government offered to the Forces' community, but the moral support the general public lent to personnel in Iraq and Afghanistan. The unpopularity of the Iraq intervention in 2003 and the lack of enthusiasm for the expanded mission in Helmand, underway from early 2006, manifested itself in the public's indifference, with a corresponding impact on military morale. Both Operations Telic and Herrick were 'wars of choice' and the Government sought to convince a sceptical public that it had made the right choice. It failed. The Military Covenant started the process by which the public came to separate the men and women from the missions and rallied to 'our boys' (and our girls), giving them unprecedented levels of moral and material support.

Following migration, civilian society's focus on the broken Military Covenant impeded any objective assessment of military performance in Iraq or Afghanistan. The shortfall in equipment resulted in part from the decision to deploy on two fronts, which undermined the success of both. Among those involved in that decision were senior commanders, who had a Covenantal obligation to the nation. The equipment available in theatre was the result of a defence acquisition process whose inadequacy cannot be blamed solely on ministers. Similarly, the Army should not shirk its share of responsibility for military accommodation, a key 'broken' Covenant issue. With the Army deploying the Military Covenant to question the government's record towards it, the concept's inherent reciprocity demands that the military should be similarly examined.

The Military Covenant introduced party politicking into defence, particularly ahead of the 2010 General Election. Following the 1998 Strategic Defence Review, the Blair government had begun to address 'people issues', on which steady progress was made until Iraq. From 2007, the Covenant provided not only the leverage for the Opposition parties to attack the Brown government for failing Britain's Armed Forces, but also the impetus for ministers to refocus their attention on the welfare of the Forces' community. Both Conservatives and Labour were eager to be seen to support the Armed Forces, but this was not manifested by a commitment to increase defence spending. Instead, in late 2009, the Defence Secretary, Bob Ainsworth, raised the possibility of introducing a legally binding Armed Forces Community Charter, which would have the effect of 'enshrining the principles of the Military Covenant in law'.[16] In 2011, the Coalition aimed to

15 Tipping, 2008, pp. 12–15.

16 Ainsworth, RUSI, 2009.

achieve the same objective with its introduction of the Armed Forces Covenant, which gave statutory recognition to the concept. In little more than a decade, the Covenant has become integral to Britain's defence policy and has provided a new prism through which to view the civil–military relationship.

Overview and Methodology

Comprising seven chapters, this study examines the genesis of the Military Covenant, its migration from Army doctrine and its entrenchment in the civilian sphere, as well as assessing its impact on Britain's civil–military relationship. More than 50 interviews were conducted for the thesis on which this book is based and it benefitted from the insights of senior officers, both serving and retired, including those involved in developing the Moral Component. Some interviewees preferred to remain off the record. Although British forces were part of a multinational coalition, this study focuses on them and the areas of operation in which they were involved: Basra rather than Baghdad, Helmand rather than Herat.

Writing in the 1970s, Capitanchik identified the 'political community' that could influence opinion and shape attitudes on defence and security matters as a limited elite in Westminster, Whitehall, Fleet Street and St James's, as well as in some academic institutions.[17] While the perspectives of former and serving defence ministers were gained, Bob Ainsworth MP, praised by many interviewees for his efforts to implement cross-government support for the Armed Forces community, was among the Defence Secretaries in office between 2004–10 who declined to be interviewed for the original work. The Royal British Legion's *Honour the Covenant* campaign was key to the concept's migration and personnel involved in it were interviewed. Other interviewees from the Forces-charity sector include the co-founder of Help for Heroes, Bryn Parry.

Evidence given to the Chilcot Inquiry and the Commons Defence Select Committee is incorporated. Other primary source material includes Hansard, military doctrine publications, the MoD's continuous attitude surveys, media reports, High Court judgments, public inquiry reports, and opinion poll data. Secondary source material includes publications and journal articles across a range of academic disciplines. Wootton Bassett was visited in 2010 as part of an assessment of the informal ceremonial that grew up surrounding military repatriation. This study primarily focuses on the period 2000–11, between the publication of *Soldiering* and the 2011 Armed Forces Act.

This study begins by investigating the Covenant's assertion that a bond of 'identity, loyalty and responsibility' between soldier, Army and nation has existed 'throughout the Army's history'. After migration, the media, politicians and charities made much of the historical nature of the Covenant. The emergence

---

17 David Capitanchik, Public Opinion and Popular Attitudes towards Defence, in Baylis (ed.), 1977, p. 261.

of the 'broken' Covenant narrative, accompanied by references to the antiquity of the concept, implied that, historically, the Covenant not only existed, but was honoured. Chapter 1 tests this claim.

Chapter 2 focuses on the development of the Moral Component of Army doctrine, particularly in the context of the Army's assertion of its right and need to be different from civilian society. British military doctrine is a living document and the paragraph encapsulating the Military Covenant was amended in 2005. In 2000, the implicit expectation is that the Covenant is non-negotiable and immutable: five years later doctrine stated: 'The nature of service is inherently unequal, soldiers may have to put in more than they receive'.[18] However, after migration, the Covenant's civilian advocates promoted the 2000 codification, with profound implications for public policy.[19]

The Military Covenant is not an agreement in any strict legal sense: there may be a moral requirement to honour its terms, but there is no legal requirement to do so. Chapter 3 explores trust-based relational or psychological agreements, comparing them with contracts that have a legal basis. Trust is at the core of all relational agreements and, according to military doctrine, is also the bedrock of operational effectiveness. The phenomenon of trust will be explored, along with the nature of covenantal contracts, in which trust is imperative. The 2005 Covenant states:

> It is a covenant, not a contract, and it is binding in every circumstance. Unless Nation, Army and soldier alike recognize and understand that it must be upheld come what may, then it fails. If it fails then first goodwill and then, ultimately, trust is withdrawn.[20]

The Covenant may be interpreted as a formalisation of the usually tacit arrangements entered into between employers and employees. However, with soldiers prepared to make the ultimate sacrifice, the consequences of failing to fulfil the Covenant are surely much greater. If more is expected from soldiers, reciprocally they should expect more from their employers, whether the Army or 'the nation'.

Chapter 4 first examines the dynamics of the Military Covenant's migration and secondly, assesses whether it narrowed what this study describes as the civil–military 'gulf'; that is, the public's indifference towards Forces' personnel involved in combat operations. The Covenant's chief advocate, General Dannatt, deployed it to marshal public support: 'The real Covenant is with the population at large – the nation'.[21] Support is considered in both the moral and material context, using metrics such as donations to Forces' charities. The chapter will also focus on

18 ADP *Land Operations*, 2005, paragraph 0715.

19 Chapter 7.

20 ADP, 2005, paragraph 0715.

21 Dannatt, London, 21 Sept 2007.

Help for Heroes, which was a catalyst for change in the Forces' charity sector. In addition, Wootton Bassett is studied in the larger context of Remembrance.

Policy-makers' reaction the post-migration Covenant and its catalytic effect in focusing their attention on Forces' community welfare are explored in Chapter 5. Government, MPs and the MoD were galvanised by the Royal British Legion's 2007 *Honour the Covenant* campaign and sought to play catch-up with the public mood. On the defensive over issues that came to be seen through the Covenant prism, the Government's response, including policy initiatives, are charted. The *Honour the Covenant* campaign is examined in depth.

Chapter 6 examines the Covenant's impact on the Army. An example of moral entrepreneurship, General Dannatt's promotion of the concept challenged longstanding civil–military relationship norms. Highlighting how the Covenant had greater utility to the Army following migration, this chapter suggests that, contingently, it became a useful diversionary tactic, deflecting attention from the conduct of the operations in Iraq and Afghanistan. The chapter assesses whether the Army is, in fact, culpable for any fracture to the Covenant, primarily in the context of the 2006 deployment to Afghanistan when the Iraq mission was far from accomplished. Civilian society extended the Covenant's scope beyond the Army; not only to the Armed Forces and the Forces' community but also to the frontline, with, for example, equipment failure or shortage becoming linked to the concept. Consequently, the Covenant was imbued with ever-greater moral significance, as it was integral to media and Opposition party narratives about soldiers' unnecessary deaths and injuries.

Following migration, different Covenants emerged as various civilian actors interpreted the concept, establishing it as 'quasi-judicial'.[22] The Haddon-Cave Inquiry into the Nimrod XV230 crash accused politicians, senior MoD officials and defence contractors of a 'systemic breach of the Military Covenant'.[23] Introducing its report on the Inquiry, BBC2's Newsnight stated:

> Remember the Military Covenant, the duty of care governments owe to the men and women prepared to risk their lives for this country? Well, the government betrayed it.[24]

Chapter 7 charts the emergence of the various Covenants and examines the extension of the concept's scope by the judiciary, not least in the 2008 High Court judgment concerning Gurkha veterans' 'right to remain' in Britain, which raised questions about the extent to which the nation values military service. Such questions go to the heart of the Covenant and underscore the significance of its impact on the civil–military relationship.

---

22 Forster, 2012b, pp. 283–300.

23 Harding, 29 Oct 2010, p. 1.

24 Jeremy Paxman, 28 Oct 2009.

Covenant Literature: An Appraisal

The Military Covenant is a recent phenomenon and academic analysis of it is sparse, especially compared with that concerning Private Military and Security Companies, organisations which have also led to reappraisals of the moral and legal issues surrounding contemporary conflict.[25] Forster pioneered Covenant analysis, with a groundbreaking paper that appeared in November 2006, just as the concept was on the brink of migration.[26] Spiked by *British Army Review*, it examines the Covenant in connection with the Army's governance. Forster paints a picture of a timid senior command, who fail to stand up to politicians and represent the Army's interests. He suggests that unless senior commanders start to give soldiers the fair treatment, value and respect that the Covenant demands, they will contribute towards breaking it. Providing the only pre-migration analysis of the Covenant and focusing on the Army and its chain of command, Forster reflects the esoteric nature of the concept at the time.

Among the changes to Britain's post-Cold War Armed Forces, which, like their American counterparts have been described as 'postmodern', is their greater permeability to civilian society.[27] Burk suggests that the US Army has to negotiate an 'ever-diminishing professional space' which is beyond its ability to control, as other agencies and organisations compete for legitimacy within its jurisdictions.[28] Using Burk's analysis as a template, Forster observes that the Covenant is a conservative concept, which attempted to reinforce the expertise, jurisdiction and legitimacy of the Army profession, while 'delegitimising alternative external conceptions of the conduct of service life'.[29] In a second paper he suggests:

> Above all, it [the Covenant] was about how a small leadership group of commanders wanted to see themselves and how everyone else should see the Army.[30]

Unlike other analysis, Forster has focused on the Covenant specifically in the context of the Army.

---

25 For example, P.W. Singer, *Corporate Warriors: The Rise of the Privatized Military Industry*, Cornell University Press, Ithaca and London, 2003; Deborah Avant, *The Market for Force: The Consequences of Privatizing Security*, Cambridge University Press, Cambridge, 2005; Christopher Kinsey, *Corporate Soldiers and International Security: The Rise of Private Military Companies*, Routledge, Abingdon, 2006.

26 Forster, 2006, pp. 1043–1057.

27 Moskos et al. (eds), 2000, pp. 1–31.

28 James Burk, Expertise, Jurisdiction and Legitimacy of the Military Profession, in Matthews and Snider (eds), 2002.

29 Forster, 2012b, pp. 283–300.

30 Forster, 2012a, pp. 273–290.

Following migration, the Army lost control of the Covenant. As McCartney observes, it had 'taken on a life of its own',[31] and according to Forster, the Covenant genie was out of the bottle. The first indication that the concept was becoming intrinsic to academic debate about wider defence issues came in November 2007 with *Out of Step*, published by Demos. Examining defence policy, the paper suggests that Armed Forces' expenditure should be better aligned to the security challenges confronting Britain; that the nation is divorced from the Forces and their needs; and that the structural changes within the management of defence had led to senior commanders losing the ability to lead and be accountable for their Services. It recommends that 'a new civil–military compact is necessary – first to restore the Military Covenant between the Army and the nation'.[32] This should be a tri-Service pledge, between the government, the military and the individual serviceman or woman, which would be a 'copper-bottom [*sic*] commitment' for life.[33] For Cornish and Dorman, problems with the Military Covenant are part of the 'organizational, bureaucratic and intellectual decay' that is endemic in Britain's defence policy.[34] They are critical of a Defence Secretary who avoided the political, financial, industrial and operational aspects of defence that were in 'a state of degeneration perhaps more serious than at any time since the end of the Cold War'[35] and instead focused on 'personnel and the issues associated with the Military Covenant'.[36]

While the media were among those in the civilian sphere who perceived the post-migration Covenant solely as a bilateral compact between government and soldiers, the concept has encouraged academics to reassess the broader civil–military relationship. McCartney suggests it is 'under strain but cannot be described as broken', observing that there are 'always inherent and unavoidable tensions' within it.[37] For Mileham, the Covenant has prompted questions not only about the Forces' *raison d'etre*, but also about whether public support can be sustained during operations and how much the United Kingdom is willing to spend on defence.[38] Tipping suggests the culpability for any Covenant 'breach' rests with Parliament – which had failed to hold successive governments to account – and with a 'hitherto apathetic public'.[39]

By 2008, the Covenant prompted policy papers from all three main political parties exploring similar themes. The Liberal Democrat's *No Choice But Change* focused on strategy as well as defence, recommending that either the defence

---

31 Ibid.; McCartney, 2010, pp. 411–428.

32 Edmunds and Forster, 2007.

33 Ibid., p. 20.

34 Cornish and Dorman, 2009, pp. 247–261.

35 Ibid.

36 Ibid.

37 McCartney, 2010, pp. 23–40.

38 Mileham, 2010, pp. 23–40.

39 Tipping, 2008, pp. 12–15.

budget be increased, or that 'our Armed Forces should do less and differently'.[40] The Conservatives' Military Covenant Commission concluded that the Covenant was in 'poor health' and 'recovery will require concerted and sustained effort by the Covenant's principal defaulter, the government'.[41] Both the *National Recognition* Report by Labour MP Quentin Davies[42] and the Commission expressed concern about the general public's lack of understanding of the Armed Forces. Among the Commission's ten-point action plan was the objective of 'normalising the Armed Forces as an integral part of national life'.[43] The Davies Report did not acknowledge the Covenant's existence, while the Commission, which cited the 2000 codification of the Covenant – rather than that of 2005 doctrine that was then current – noted that 'it is drafted in vague terms that limit its usefulness'.[44] However, it was this very vagueness that gave the Covenant the malleability that had proven so useful: the Coalition government ran into difficulty when it tried to make this amorphous concept concrete in statute.

The contractual nature of the Military Covenant has provided an additional dimension to the analysis of the legal as well as the moral issues affecting the Armed Forces. Noting the concept's plasticity, Forster argues it has become the 'key vehicle' for holding the MoD to account in its duty of care to Service personnel.[45] It has also drawn renewed attention to the absence of legal contract in connection with employment in the Armed Forces. This is tangential to analysis but merits further scrutiny, not least because of the increasing juridification of the Armed Forces, examined by Rubin and Forster.[46] McCartney describes the Covenant as a 'moral' contract; for Tipping, it is an 'implicit' contract and one that is closest to the 'psychological contract'.[47] Mileham suggests that in the mid-1990s this contract was investigated by the 'Army Department but was discarded in favour of the Military Covenant'.[48] His investigation of the Covenant – or, more accurately, *Soldiering: The Military Covenant* – is undertaken in the context of the Army's institutional ethics and what Hackett described as soldiers' 'unlimited liability'.[49] After 2010, the Coalition's determination to enshrine the Military Covenant in law – which drew an unenthusiastic response from Ingham and Dandeker[50] – resulted in the anaemic Armed Forces Covenant, following the government's realisation that it was entering a legal minefield. However, Tipping

40 Campbell, 2008, p. 7.

41 Conservative Party, 2008, p. 4.

42 Davies, 2008, p. 4.

43 Conservative Party, 2008, p. 8 (an ambiguous aspiration, as visitors to Buenos Aires are told).

44 Ibid., p. 9.

45 Forster, 2012b, pp. 283–300.

46 Rubin, 2002, pp. 36–57; Forster, 2012b.

47 Tipping, 2008, pp. 12–15; McCartney, 2010, pp. 411–428.

48 Mileham, 2010, pp. 23–40.

49 Hackett, 1983, p. 72.

50 Ingham and Dandeker, 2010.

had warned in 2008 that the Covenant's inherent expectations were 'impractical to describe in anything other than generalisations': any attempt to introduce legal sanction if they went unfulfilled was a 'bureaucratic impossibility'.[51] Highlighting the legally anomalous status of Armed Forces personnel, Mileham suggests the concept of unlimited liability was dropped from Military doctrine, because 'deeper considerations about language suggested that "liability" is too legalistic a term'.[52] For McCartney, the unlimited liability taken on by soldiers is the 'key element of the bargain embodied in the Military Covenant'.[53] Academic analysis reflects the tensions in trying reconcile the Armed Forces and civilian law, skirmishes in which the Forces are invariably wounded. The 'legal component', as identified by this study, is a background theme of this work.

Literature on the Covenant is, to date, generally limited to journal articles. One exception is a work by the Commission's convener, Andrew Murrison MP, which includes a foreword by the Prime Minister, David Cameron. It provides an insight into Conservative thinking and Coalition policy development. For Murrison, 'there is nothing new about the Military Covenant': he designates Elizabeth I the first 'Hero of the Military Covenant' for introducing the 1593 Act for the Relief of Soldiers.[54]

The foundational text for many explorations of the civil–military relationship is von Clausewitz's *On War*. With Clausewitz's identification of the Trinitarian aspect of conflict – the interaction of passion, chance and reason, evinced by a secondary trinity of people, military and government – inevitably it has been inferred, at least by some interviewees for this study, that he provided a template for the Military Covenant. The Clausewitzian trinities were expected to be in a state of flux and 'variable in their relationship to one another', but their interdependence was emphasised: it was imperative to attain 'a balance between these three tendencies'.[55] General Dannatt considered that the balance between the Covenant's trinity had been lost, reflected by his statement it was 'out of kilter'.[56] Somewhat unexpectedly, the author says that he was uninfluenced by Clausewitz, not having read him when he wrote *Soldiering: The Military Covenant*.[57]

The continuing relevance of *On War* – which Strachan reminds us is a work in progress whose judgements are not consistent – has repeatedly been questioned, not least by Liddell Hart who found Clausewitz's ponderous tones so solid as to cause mental indigestion.[58] An investigation into the theory and nature of conflict, the statement that war is an alternative political instrument has become a truism

51 Tipping, 2008, pp. 12–15.
52 Mileham, 2010, pp. 23–40.
53 McCartney, 2010, pp. 411–428.
54 Murrison, 2011, p. 9.
55 von Clausewitz (eds and trans. Howard and Paret, introduction Heuser), 2007, p. 31.
56 Rayment, 11 Nov 2007.
57 Roberts, interview.
58 Liddell Hart, cited by Strachan, 2007, p. 16.

in the last two centuries. Clausewitz is challenged today by, among others, van Creveld, who has argued that an early nineteenth-century state-centric doctrine of warfare 'waged by the state, for the state and against the state' is far from the best way to understand either 'uncivilized' war or the great wars of the twentieth century.[59] Critics cite the work's failure to address economic and naval warfare, as well as the scant attention given to unconventional war and insurgency, which is likened to 'smouldering embers'.[60] However, Smith, who identified 'war amongst the people', argues that 'the trinity is crucial to all forms of war, to this very day'.[61] Clausewitz's many detractors, including Luttwak and Keegan, suggest that the change in the nature of war, along with the technology with which it is fought, has rendered his theories obsolete: Metz argues that it is time for strategists to hold a wake and move on.[62] However, Waldman observes that although many have attempted to breach the Clausewitzian edifice, 'few have launched an all-out offensive in order to impose their own distinctive general theory of war'.[63]

Today's analysis of the civil–military relationship in Britain has been shaped by concepts that, for the most part, emanated from the United States in the early Cold War years. *The Professional Soldier*, *The Man on Horseback* and *The Soldier and the State* are seminal works, described by Forster as 'foundational texts' that set the scene for subsequent research.[64] They are publications of their time, when vestigial suspicion about standing armies was manifesting itself in concerns about the military-industrial complex. They address the 'civil–military problematique' identified by Feaver: how a democratic government best controls its military without compromising the military's functionality or society's ideals and institutions.[65] Huntington identified an ideal of 'objective civilian control' based upon the recognition of military professionalism, key to which was its officer class, imbued with the idea of service to the nation.[66] He might have disdained the inclusivity of the Military Covenant, which speaks to 'each individual soldier': the rank-and-file are specialists in the application, not the management, of violence; 'their vocation is a trade not a profession'.[67] In his sociological study of the American military, Janowitz warned of the dangers of 'unanticipated militarism' engendered by the growth of a military establishment since 1939, which in contrast to its British counterpart, was perceived to have too much influence over legislation and public opinion.[68] The constabulary force advocated by Janowitz resolves this problem

59 van Creveld, 1991, p. 36.
60 von Clausewitz, 2007, p. 185.
61 Smith, 2005, p. 58.
62 Metz, 1994, p. 126.
63 Waldman, 2009.
64 Janowitz, 1962; Huntington, 1957; Finer (who was British), 1962; Forster, 2005, p. 20.
65 Feaver, 1996, pp. 149–178.
66 Huntington, 1957, p. 94 and p. 35.
67 Ibid., p. 18.
68 Janowitz, p. 14.

because it is 'designed to be compatible with the traditional goals of democratic political control', in large part because of the professionalism, self-esteem and moral worth of the officer class which has a 'meaningful interaction with civilian values'.[69] For Caforio, this constabulary force represented a convergence of civilian and military spheres of interest and activity, with the military continuously having to adapt to an ever-changing society and incorporating its values: in contrast Huntington presents a divergent model of the two spheres.[70]

The foundational texts are historical documents, providing an insight into the civil–military preoccupations of the early Cold War era, a time of decolonisation and military *coups d'etat*. Finer observed that military intervention in politics in post-colonial successor states tended to be the rule. His taxonomy of military intervention – supplantment, displacement, blackmail, influence – depended upon the prevailing political culture, identified as minimal, low, developed or mature.[71] Despite Britain's highly developed political culture, Finer suggests it had been far from immune from intervention by the military during the twentieth century, although the intervention level ranged from influence to blackmail rather than anything more malign. Displacement let alone supplantment by the military of the civil power in Britain today is so far-fetched as to be risible: however, Strachan suggests that the accepted norm that the British Army is apolitical does not stand up to investigation. What limits the impact of intervention is the political culture within which the Army operates, rather than that Army itself eschewing politics.[72] With defence having a higher priority in policy-making across the Atlantic, it is unsurprising that the civil–military relationship is the focus of more academic analysis. However, recent social change has engendered further academic approaches to civil–military relations, with the equal opportunities agenda and the impact of diversity upon the ethos of the Armed Forces becoming the subject of inquiry on both sides of the Atlantic,[73] while the post-Cold War pan-European transition to volunteer forces is also an area of examination.[74]

The declaration of the War on Terror and the 2003 Iraq intervention gave new impetus to civil–military relations analysis in the United States, concerning the covert and overt influence of the military on policy-making and the issue of civilian authority over the military.[75] With a budget of $670 billion in 2010, Federal spending on defence exceeded that on education, making Eisenhower's

69 Ibid., p. 440.

70 Caforio, p. 18 and p. 16.

71 Finer (Introduction Stanley), 2002, 1962.

72 Strachan, 1997, p. 19.

73 For example, Trish Winter and Rachel Woodward, 2007, *Sexing the Soldier: The Politics of Gender and the Contemporary British Army*, Routledge.

74 Tresch and Leupracht (eds), 2010.

75 E.g. Andrew W. Bacevich, 2005, *The New American Militarism: How Americans Are Seduced by War*, Oxford University Press Inc, USA; Richard B. Myers and Richard H. Kohn, 2007, Salute and Disobey, *Foreign Affairs*, Sept/Oct.

concerns about an over-mighty military-industrial complex not just the stuff of febrile conspiracy theories. Former diplomat George Kennan observed that a military intervention such as Somalia 'is something that the Founding Fathers of this country never envisaged or would ever have approved'. A critic of the Iraq conflict, Kennan was critical of 'the extreme militarization not only of our thought but our lives'. It seems that Lasswell's concerns about a 'garrison state' were not so fanciful. Feaver's analysis of the US civil–military relationship in the context of the principal–agent relationship suggests that the post-war US military – the 'armed servants' of the state – frequently 'shirked' rather than worked: i.e. did not act to the principal's satisfaction.[76] Defense Secretary Rumsfeld's reminder that 'the Constitution calls for civilian control of this Department' was an indication of civil–military tensions within the Bush administration after 2003, which found its way into analysis that, while it did not necessarily invoke the foundational texts, conveyed a similar sense that the US military was in danger of becoming over-mighty.[77] With Rumsfeld, the 8,000-mile screwdriver, imposing his 'war-lite' transformation agenda on a military institutionally still attached to the Powell Doctrine of dollar-heavy overwhelming force, a civil–military clash was inevitable, especially from 2004 when the US media began talking about 'Saigon on the Tigris'. In the post-Cold War era, senior commanders have strayed out of their politically neutral lane, swerving into policy-making (Powell over Bosnia), partisanship (overt Republican support) and insubordination (McChrystal), from which it can be inferred there is frequently dysfunction within upper reaches of the civil–military relationship.

In the late 1990s the 'civil–military gap' – another conceptual framework emanating in the United States – became a major theme in analysis of the relationship between the Armed Forces and society. Gap analysis focuses on the divide between the two spheres. Feaver and Kohn led a major study under the auspices of the Triangle Institute of Security Studies:[78] based on surveys of opinions among elite policy-makers and military officers, it focused on foreign and defence policy-making, civil–military relations and cultural values.[79] Feaver identifies a necessary 'functional' gap between the military and civilian spheres and a 'troublesome' gap of distrust, misunderstanding and possible alienation.[80] The study revealed a growing divergence between the two realms in connection with the willingness of the military to subordinate itself to civilian political control and the cultural divide between a self-sacrificing military and the selfish individualism of civilian society. Kohn argues the gap might have become 'so wide that it threatens the effectiveness of the Armed Forces and civil–military

76 Feaver, 2003a.

77 Desch, 2007, pp. 97–108.

78 Feaver and Kohn (eds), 2001.

79 Feaver, 2003b, pp. 1–5.

80 Ibid.

cooperation'.[81] However, the gap is multi-dimensional and dynamic: within the norm of civilian dominance, the civil and military worlds are constantly renegotiating their relationship. Burk advances an analysis of the gap based upon the Armed Forces' 'institutional presence' in society, assessed through their claim to material resources and their moral integration.[82] Roman and Tarr highlight the civil–military convergence among the elites involved in national security policy-making, while Bianco and Markham's analysis of the declining military experience of Congress emphasises divergence.[83]

Following the example of civil–military paradigms and taxonomies set in the United States has often led to a conceptual cul-de-sac on this side of the Atlantic, where questions about military subordination to civilian control have been settled since 1689. Since the Glorious Revolution, the military has been *a part of* but *apart from* civilian life. The two spheres have overlapped on an institutional level through the Crown. Inevitably, there has been a civil–military overlap at an individual elite level particularly during the eighteenth century and the early nineteenth-century era of the *ton*, when an upper 10,000 dominated British institutional, political and cultural life. However, the civilian and military worlds have remained generally distinct, not least because conscription is the historic exception rather than the rule. In additional, the Army has usually been comparatively small and, for much of its time, stationed outside the United Kingdom. As Strachan suggests, in Britain, the civil–military gap is 'a product of continuity'.[84] Similarly, this study assumes that a civil–military 'gap' has been, and remains, the norm in Britain, but will examine whether a civil–military 'gulf' developed after the Iraq intervention of 2003 and whether the Covenant contributed towards narrowing it.

Gap analysis is integral to assessments of the Military Covenant, tending to focus on the cultural divergence between the Armed Forces and society. Tipping, like others, looks to the World Wars when there was inevitably greater empathy between the two spheres. She suggests that while today's Servicemen and women make comparable sacrifices – from foregoing some civil rights to risking health and life – this is not reciprocated in terms of civilian society's 'support', something McCartney identifies as what the military wants above all else from the public.[85] Tipping acknowledges that the relationship between the British public and the military has been 'uneasy'.[86] For McCartney, the military has been respected rather

81 Peter D. Feaver, Richard H. Kohn and Lindsay P. Cohn, The Gap Between Military and Civilian in the United States in Perspective, in (eds) Feaver and Kohn, 2001, p. 1.

82 James Burk, The Military's Presence in American Society, ibid.

83 Peter J. Roman and David W. Tarr, Military Professionalism and Policymaking: Is there a Civil–Military Gap at the Top? If So, Does it Matter?, ibid. William T. Bianco and Jamie Markham, Vanishing Veterans: The Decline of Military Experience in the US Congress, ibid.

84 Strachan, 2003, pp. 43–63.

85 Tipping, 2008, pp. 12–15 and McCartney, 2010, pp. 23–40.

86 Tipping, 2008.

than understood. She is concerned that the Covenant itself invokes 'an idealized past' where relations between the Military and wider society were close, with the implication that, historically, it was upheld.[87]

Conclusion

While imported analysis encourages reappraisals of the British civil–military relationship, its agenda is, in general, set in the United States, primarily for the United States. Until recently, analysis in Britain can only follow in the American academic slipstream. However, the evolving Military Covenant provides a uniquely British perspective. This study suggests that, in addition to being a useful conceptual tool with which to analyse the Army's relationship with its soldiers and the civilian world, the Military Covenant now forms the dominant paradigm for any investigation into civil–military relations in twenty-first-century Britain.

87 McCartney, 2010.

Chapter 1
The Historical Context

Introduction

The archaic quality of the word 'Covenant' is reinforced by its Biblical associations. Similarly, the term Military Covenant implies that the concept has its roots deep in the past; a compact, perhaps written on parchment with a plumed quill, familiar to the Iron Duke or Cromwell's Ironsides. However, barely in its teens, the Covenant is an 'invented tradition', in which the Army itself is steeped.[1] Particularly in the wake of Army reorganisations, many regimental authorities have 'mobilized the past in the service of the present and the future', not least to engender a sense of *esprit de corps*.[2] Members of the five separate regular battalions of the Royal Regiment of Scotland, created in 2004, still wear distinctive coloured hackles: the so-called Government tartan adopted by the Regiment was previously worn by the Black Watch and the Argyll and Southern Highlanders. However, the tartan kilt itself – emblematic of the Scottish Highlands – was the invention of an English industrialist in the 1720s.

The Military Covenant was invented in 2000, in a background essay that provided the 'philosophical underpinning' to the Moral Component of Army doctrine.[3] *Soldiering: The Military Covenant* aimed to convey the truths about being a soldier today, in particular the moral responsibilities concerning the possible taking of life.[4] The Covenant concept, an aspiration based on the principle of reciprocal obligation between soldier, Army and nation, was one aspect of *Soldiering*.[5] As Hobsbawm suggests, invented traditions 'normally attempt to establish continuity with a suitable historic past'.[6] Given the importance of history and tradition to the Army as an institution, reflected by, for example, the reverence that has always been accorded to colours and trophies,[7] no exploration

---

1 Hobsbawm and Ranger (eds), 1989.

2 French, 2004, p. 84.

3 Roberts, interview.

4 Ibid.

5 Paragraph 103, ADP5, 2000a. Army Doctrine Publications. Ministry of Defence. © Crown copyright 2014.

6 Hobsbawm, in Hobsbawm and Ranger (eds), p. 1.

7 In the 1840s, Chaplain-General Gleig criticised the neglect of colours captured by Marlborough that had been given to St Pauls Cathedral: their staves were being used to chase away rats.

of contemporary soldiering can be unmindful of the past. However, from 2006 and its migration from Army doctrine, the Covenant became a trope to convey the current difficulties faced by an Army war-fighting on two fronts. This study suggests that the Covenant's successful entrenchment in the civilian sphere can in part be attributed to erroneous assumptions concerning its antiquity. The iteration of Paragraph 103, which came to be cited – in whole or in part – in Parliamentary debates, press reports, a High Court judgment, public inquiry reports and academic articles, reinforced the assertion that a centuries-old 'unbreakable common bond' of 'loyalty, identity and responsibility' had, in fact, existed. With the exception of McCartney, herself an historian, few civilians have challenged this claim.[8] Indeed, two of the Covenant's leading civilian advocates – the Royal British Legion and the Conservative Party – have emphasised its sixteenth-century roots. The co-option of the past in Army doctrine lent the Military Covenant gravitas and the patina of authenticity, just as the incorporation of the hackle and kilt in its uniform provides a reflexive 'Scottishness' to Royal Regiment of Scotland. Civilian society's assumptions about the concept's antique provenance contributed to the Covenant's success, underlined by the legitimisation in statute of this invented tradition.

This chapter examines whether the Military Covenant simply codified an unwritten centuries-old bond between the individual soldier, Army and nation (both policy-makers and public), or whether claims concerning such a bond have been idealised or exaggerated. To this end, evidence of the historical existence of a reciprocal sense of identity, loyalty and responsibility between the Covenant's three component actors – soldier, Army and nation – is examined. The bilateral relationships within the Covenant's tri-lateral framework are explored; that is, between the individual soldier and the nation, the nation and the Army, the Army and the individual soldier. In addition, there will be a certain degree of synergy between the qualities of identity, loyalty and responsibility: the generation of one could stimulate the growth of another.

Today's Army began to emerge following Charles II's restoration in 1660. Given this chapter's brevity, the history is selected rather than detailed; concentrating on regular rather than auxiliary forces, as well as those that came to be funded by the annual Army Estimates bill rather than those that, until 1858, were supported by the East India Company. The Army will be focused upon, rather than the regimental system, 'the principle [*sic*] vehicle of the nation's military culture'.[9] Parliaments voted money to the Army, after 1689 the annual Mutiny Acts gave the Army legitimacy, while the individual soldier was subject to a body of military law. Since migration, various matters have been deemed 'Covenant issues' by different civilian actors: this chapter focuses primarily on those having a direct impact on soldiers' welfare: pensions and healthcare. Remembrance is assessed in Chapter 4. Whether its claims to historical authenticity are justified, the Military

8 McCartney, 2010, pp. 411–428.

9 John Keegan, in French, 2004, p. 1.

Covenant provides an additional prism through which to view the civil–military relationship in the past.

The Bond of Identity

The American War of Independence (1775–83) was considered by some contemporaries in Britain to be a civil war. One writer stated the Americans had 'the manners, habits, and ideas of Britons'; they had in common 'the same laws, the same religion, the same constitution, the same feelings, sentiments'.[10] In short, the Americans and British shared many of the civic and cultural institutions, norms and beliefs that Smith identifies as contributing to a sense of cohesive national identity.[11] However, Tom Paine argued that America was entirely different, not least because of the multi-national nature of colonial society in the 1770s.[12] Identity is elusive to capture, defined at its simplest as our understanding of who we are, either individually or collectively. For Muir and Wetherell, identity is a personal matter but a collective force, while 'identifying with something always requires the exercise of individual agency'.[13] Some American colonists believed they were British – itself a fairly recent construct – others did not. This mercurial aspect to identity, whether individual or collective, derives from the fact that what it tries to pin down is rarely static and it involves subjective interpretation. Depending on the identity of the audience, the identity of 'self' that the individual presents can be changed at will, as Goffman describes.[14] Equally, a sense of collective national identity can emerge from, and be reinforced by, an 'other'. Created *de jure* by the Act of Union in 1707, the British nation was, according to Colley, 'an invention forged by war'.[15]

A sense of British identity and allegiance to Britain was engendered by more than a century of intermittent conflict, primarily against France, ending at Waterloo in 1815. Catholic, absolutist, then revolutionary, it was the 'other' just across the Channel. In addition, the spoils of the eighteenth-century wars – overseas territory – augmented British formal and informal control over increasingly large parts of the world. Between 1698 and 1815, Britain was involved in the Nine Years War, the War of Spanish Succession, the Wars of Jenkins's Ear and Austrian Succession, the Seven Years War, the American War of Independence and then finally, the wars against Revolutionary and Napoleonic France. As Bowen observes, 'war was a semi-permanent feature on the historical landscape'.[16]

10 Cited by Wahrman, 2001, pp. 1236–1263.
11 Anthony D. Smith, 1991, p. 101.
12 Wahrman, 2001.
13 Muir and Wetherell, 2007, p. 10.
14 Goffman, 1990.
15 Colley, 1992, p. 5.
16 Bowen, 1988, p. 4.

The *British* Empire – essentially a collection of far-flung territories amassed piecemeal from the sixteenth century – became a source of British identity, cohesion and pride.[17] How far successive governments actively pursued or nurtured imperial expansion is a matter of historiographical debate. According to Baugh: 'England began her career as the greatest and most prosperous colonizing power that the world had ever known without any fixed policy, in fact without any clear idea of what she and her people were doing'.[18] However, by 1815, Britain had 43 colonies on five continents.[19] For an 'island nation', the collective British outlook was, paradoxically, outward looking, not insular.

Warfare demands the public's more active identification, and engagement, with the nation. Brewer highlights how, during the eighteenth century, Britain evolved into a 'fiscal-military state' as conflict necessitated a wholesale change in governmental administration and financial practice.[20] Manpower requirements spiralled: the Army averaged 93,000 men during the War of the Spanish Succession (1702–13), peaking to 233,853 in 1815.[21] As significantly, there was a steady rise in the proportion of the population who undertook some form of military service in the auxiliary forces, from one in 16 during the War of Austrian Succession (1740–48), to one in five or six during the Napoleonic Wars.[22] After the 1756 invasion scare, Gibbon observes: 'The country gentlemen of England unanimously demanded the establishment of a militia enabling them to do their patriot duty'.[23] Similarly, with the invasion threats between 1779–82 and in 1803, an estimated half a million men were drawn into civil defence, with society mobilised on a scale not seen until the First World War. Colley observes that 'training in arms under the auspices of the state that was the most common collective working-class experience in the late eighteenth and early nineteenth centuries'.[24]

Although the British identity was brought about by war, there was little identification with regular soldiers by civilian society for almost the first 200 years of the Army's existence. Despite the military victories that began to deliver an overseas Empire and civilian involvement in the auxiliary forces raised for national defence, until the First World War, the Royal Navy was a greater source of national identity, dovetailing with Britain's view of itself as an island nation and imperial hegemon. The over-arching 'blue-water policy', explored by Baugh among others,[25] had led British strategy favouring colonial and mercantile concerns over

17 Colley, 1992, p. 132.

18 Daniel A. Baugh, Maritime Strength and Atlantic Commerce, in Stone (ed.), 1994, p. 189.

19 Ferguson, 2001, p. 73.

20 Brewer, 1989, p. 22.

21 Holmes, 2002, p. 134.

22 Bowen, 1988, p. 14.

23 Gibbon (1984 edition), p. 122.

24 Colley, 1992, p. 312.

25 Baugh, 1998.

engagement in Europe. The senior Service protecting the national interest, whether deterring invasion at home with a 'wooden wall' or facilitating trade overseas, the Navy also provided Britain with a 'contemporary heroic culture' from the mid-eighteenth century, not least with Vernon's victory at Portobello that inspired '*Rule Britannia*'.[26] In 1796, Cobbett framed national identity in terms of the Navy, despite the oceans, fleet and sailors being an 'imagined community' for many Britons from rural areas.[27] As Brewer suggests: 'The great military buildings of eighteenth century England were not barracks and forts but the dry-docks, stores, roperies and building yards of the Royal Navy'.[28] For Ruger, until the Edwardian era, the Navy was one of the most important 'agents of Britishness'.[29]

The Military Covenant's assertion of an unbreakable bond of identity between soldier, Army and nation is a misconception until the Crimean War and somewhat of an exaggeration after it. Until the latter half of the nineteenth century, civilian society regarded regular soldiers as 'other', a potentially dangerous underclass. Chronic manpower shortages led to the 1695 Act for the Relief of Poor Prisoners for Debt or Damages, which allowed prisoners their pardon in exchange for military service: similar Acts would follow throughout the eighteenth century, as they had in the past. During the Elizabethan era, Newgate prisoners were freed to supplement expeditionary forces: in 1597, the Privy Council ordered that all the able-bodied unemployed in Surrey and Sussex were to be sent to the Netherlands.[30] Impressment, which continued sporadically until 1780, was a convenient method of ridding localities of ne'er-do-wells.[31] Before barracks were built in 1790s, civilians regarded soldiers as unwelcome billetees; afterwards as the inhabitants of a 'total institution' who, until flogging was abolished in 1881, were subject to inhuman discipline. The same year, the commanding officer of the 2/18th Foot stated that 10 per cent of his soldiers were members of the 'criminal class' and more than one third were illiterate. The military's difference from civilian society was underlined by the exclusion of women, with a maximum of seven per cent of soldiers allowed to marry 'on the strength' until the 1870s. In 1796 Lord Erskine denounced the 'uncontrollable licentiousness of a brutal and insolent soldiery'.[32] With one-third of all admissions to hospital in Aldershot in 1861 related to venereal disease and 10,966 courts martial for drunkenness in 1868, it seems that three quarters of a century later his lordship still had a point.[33] Despite British troops

26 Jenks, 2006, p. 24.
27 Ibid., p. 2.
28 Brewer, 1989, p. 24.
29 Ruger, 2004, pp. 159–188.
30 Act of Privy Council 27/290 in Cruikshank.
31 Cruikshank, p. 94.
32 Brereton, 1886, p. 43.
33 French, 2004, p. 109 and p. 113.

being victorious in at least 15 small wars between 1815–54, such was soldiers' reputation that many considered themselves disgraced if a son enlisted.[34]

The collective wariness evinced by British society towards soldiers was not reflected by its cultural landscape, in which feats of arms were celebrated and idealised. Coutu observes that after 1758, an 'overtly triumphalist' culture grew up in connection with imperial conquest.[35] An increasing number of monuments in public space came to be commissioned by order of Parliament, rather than funded by private donations. Westminster Abbey, the most visited place in Britain, became an 'indoor Elysium',[36] crowded with statues depicting fallen heroes such as Wolfe and 'enchained defeated foes', reminiscent of Roman iconography of imperial war.[37] The print of Benjamin West's *Death of General Wolfe* exhibited in 1771 was to be outsold by an equally fictitious depiction, *Marquess of Granby Relieving a Sick Soldier* by Edward Penny.[38] Published in 1817, *The Burial of Sir John Moore after Corunna* entered the canon in Britain and the United States.[39] Historians might be divided on his baleful or beneficent hold on the Army, but Wellington himself remained a national hero long after his death in 1852, with Waterloo Day a celebrated part of the national calendar.[40]

For the century after Waterloo, the Army's *raison d'être* was primarily 'the preservation and expansion of Empire',[41] which offered it a 'perpetual battlefield'.[42] Queen Victoria stated: 'If we are to maintain our position as a first-rate Power, we must, with our Indian Empire and large Colonies, be prepared for attacks and wars, somewhere or other, continually'.[43] Attacks and wars finally benefitted soldiers in terms of civilian society's approbation after Crimea, as they were identified as instruments of beneficent British rule, coming to be imbued by the public with a sense of moral, rather than purely military, mission. As Said suggests, the British prided themselves on the moral dimension of their imperial power, with their focus being on the 'great legitimizing narratives of emancipation and enlightenment'.[44]

---

34 An 1867 Royal Commission reported that mothers would prefer a son's death to his enlistment (Samuel Haden, Secretary of the Army and Navy Pensioners Employment Society).

35 Coutu, in Bonehill and Quilley (eds), 2005.

36 Ibid., p. 44.

37 Craske, in Bonehill and Quilley, p. 47.

38 Alan J. Guy, The Army of The Georges 1714–1783, in Chandler and Beckett (eds), p. 241.

39 Robson, 2009. (With thousands of soldiers interred in unmarked graves during the Napoleonic Wars, the poem surely resonated with the bereaved.)

40 Strachan (1984) questions the orthodoxy that a 'long round of post-Waterloo sleep ends in a rough awakening in the Crimea and a round of reform follows'.

41 Strawson, 1991, p. 195.

42 James Morris, *Heavens Command*, Faber and Faber, 1973, cited by Strawson, p. 136.

43 Ibid.

44 Said, 1994, p. 186.

Soldiers brought not just civilisation, but Christian civilisation, to the wider world. How far they considered themselves the military wing of the Missionary Society is open to question, but as an institution the Army recreated itself as 'an almost ostentatiously Christian Army', not least during the Chaplain Generalship of Gleig and the growth of evangelism.[45] The Army's image as a 'moral wasteland' and 'shelter for black sheep of various breeds' came to be transformed.[46] The public's attitude towards soldiers grew more favourable with the rise of temperance. From 1863, like other regimental journals, the 12th Foot's *East Suffolk Gazette* reports on meetings of the regiment's Total Abstinence Society. Generals Havelock and Gordon were perceived as Christian heroes, with both the Indian Mutiny (1857) and the Siege of Khartoum (1885) reinforcing the British sense of moral rectitude when contrasted with the Moslem, Sikh or Hindu 'other'. Towards the end of the nineteenth century, French suggests that 'the soldier in the abstract had become an icon held in growing public esteem. His character was construed as being loyal, patriotic, brave and therefore virtuous'.[47]

As they garrisoned the Empire throughout the nineteenth century, British soldiers could be likened to the convicts transported to Australia until 1863: far across the oceans, out of sight if not completely out of mind, and maintained ideally at minimal cost to the taxpayer. Expenditure on the Army and ordnance fell from £43m in 1815 to £10.6m in 1820, dropping to below £8m in 1835. Although the penetration of an imperial sensibility among the British at home has been the subject of academic inquiry,[48] Said suggests that Kipling – creator of the archetypal imperial soldier, Tommy Atkins – reflected how it was 'uncontested'.[49] One by-product of the improved literacy after the Foster Education Act of 1870 was the expansion of the popular press, which 'promoted the Army as instruments of Imperial mission'.[50] The battlefront was no longer solely depicted through the high cultural media of oil paint and marble: from the 1860s, newspapers could utilise the telegraph and photography, bringing the battlefront to the home front with unprecedented immediacy. Conan Doyle and Churchill were among the reporters sent to the Cape.[51] According to one contemporary: 'Probably England never waged a war which excited so keen an interest as the great Boer War'.[52] The pervasiveness of militarism throughout the Victorian era is reflected by 'the cultural products of nationalism', whether poetry, fiction, music and the visual arts.[53]

---

45 Olive Anderson, 1971, pp. 46–72.

46 MacKenzie (ed.), p. 22 and Anderson, 1971.

47 French, 2004, p. 233.

48 E.g. Bernard Porter, *The Absent-Minded Imperialists: Empire, Society and Culture in Britain*, Oxford University Press, Oxford, 2006.

49 Said, 1994, p. 162.

50 Spiers, 1980, p. 213.

51 Ibid., p. 238. (They were among the 20 correspondents sent by *The Times*.)

52 Hardy, 1900, p. 25.

53 Olive Anderson, 2006, p. 141.

Popular militarism was an unremarkable part of life – whether Fleet reviews or in boys' comics and books.[54]

The bond of identity between Army, soldier and nation should have been strengthened with repeated deliberate attempts to bind the Army with distinct localities. The 2004 reorganisation of the Infantry attempted to emulate the spirit of the 1870s Cardwell-Childers reforms by boosting regimental links with communities, not least to help soldiers reintegrate back into civilian life.[55] Local linkage was also a characteristic of the Haldane reforms after 1906, which restructured the auxiliary forces, seen as another bridge between the civilian and military spheres. However, the schism between regulars and the auxiliaries – who they frequently disparage as weekend warriors – is often only bridged by shared combat experience.[56] According to French: 'Only a minority of regiments struck such deep roots into their local communities that they could recruit the majority of their men from their own districts'.[57] Despite a public outcry following almost every restructuring of the Army – especially with the 'loss' of regiments with county names – the policy of local linkage was never properly fulfilled.

Civilian society's sense of identity with the Armed Forces was probably at its zenith during the World Wars and the introduction of conscription. However, as French observes, while Regular officers and long-service NCOs had 'an almost metaphysical vision of what constituted their regiment', conscripts had a different understanding of the military community, one that was based on the camaraderie of shared hardship.[58] Some 2.5 million men went through National Service between 1945 and 1963. With more than 1,000 fatalities in Korea, many did far more than 'peel spuds in Catterick'.[59] Whether this closure of the civil–military 'gap' is such a positive for the Army is seldom considered. During the 1950s, National Service bred, if not contempt, a lack of reverence for the Armed Forces reflected in popular culture, in films such as *The League of Gentleman* or the TV sitcom, the *Army Game* (1957–61). As Cowper-Coles observes, today's attitudes towards the military are 'more deferential and less balanced'. If the media reflect their consumers, today the British appear happy to identify with and support 'our boys', in part because they have an idealised view of soldiers, due their lack of familiarity with the Armed Forces.[60]

Commanders frequently shared the nation's lack of identity with the rank and file. Wellington's mistrust was such that he refused to countenance the abolition of flogging. Captain Richard Pope of Schomberg's Horse wrote: 'Such a set of

54 Jeffrey Richards, Popular Imperialism and the Image of the Army in Juvenile Literature, in MacKenzie (ed.), 1992.

55 Dorman, 2006, pp. 489–502.

56 Duke of Westminster, interview.

57 French, 2004, p. 58.

58 French, 2004, p. 283.

59 Former National Serviceman.

60 Chapter 5.

ruffians and imbeciles you never beheld, you may call them cannon fodder but never soldiers'.[61] An absence of sympathy between officers and men would be unsurprising: Towle stresses the social gulf that existed until the mid-nineteenth century between the classically educated elite and the semi-literate or illiterate masses.[62] As a staff sergeant stated in 1846: 'The British soldier is a neglected man ... a pariah of the body politic ... His own officers despise him and the public at large despise him'.[63] Victorian-era regimental gazettes reflect an iron divide between officers and men: whether at cricket, concerts, the turf, polo, theatricals, or pig-sticking, the two sides never mixed.[64] Until the aftermath of the Crimean War, the welfare and living conditions of the rank-and-file appear for the most part to be of little consequence to senior officers, who personified 'Horse Guards' and an 'unyielding traditionalism'.[65] In addition, officered by members of the upper classes who commanded those perceived to be the society's dregs, the Army was hardly representative of a Britain that from the late eighteenth century became increasingly industrial and urban. It was out of step with a growing number of 'respectable' members of the new working class: literate, religiously non-conformist and – although not yet enfranchised – politically engaged.[66] Among the Cardwell reforms of the 1870s was the abolition of Purchase. Spiers suggests that despite the aim to broaden the social mix of the officer corps, little actually changed: 'Private means remained the principal requirement for a commission in the home Army'.[67] National Serviceman and journalist Anthony Howard observed that in the 1950s a commission required an acceptable accent: 'It was social apartheid from the beginning'.[68]

A contract of employment, if one had existed between soldiers and the Army, would have included an implicit 'unlimited liability' clause.[69] Given the poor pay, low status, harsh living conditions and draconian discipline they had to endure until after Crimea, soldiers – especially the 2,217 who were flogged in 1855[70] – might have wondered whether the Army was fulfilling its side of the bargain. However, the separate code of military justice ensured that until the 1881 Army Act, a soldier's conduct was minutely regulated, to the extent that he could have ceded any moral agency or responsibility. However, French points out that for most soldiers, the Army was a 'vague legal abstraction':[71] soldiers identified

61 Brereton, 1886, p. 25.

62 Towle, 2009, p. 9.

63 Sergeant J.M. MacMullen cited by Burroughs, 1985, pp. 545–571.

64 For example, *Journal of the Household Brigade*, *Rifle Brigade Chronicle*, *St George's Gazette*, *Bengal Tiger*.

65 Burroughs, 1985, pp. 545–571.

66 Thompson, 1991.

67 Spiers, 1980, p. 249.

68 Hickman, 2004, p. 58.

69 Hackett, 1983, p. 73.

70 Burroughs, 1985, pp. 545–571.

71 French, 2004, p. 5.

themselves primarily with their regiment, 'the keystone of our military arch'.[72] Wolseley stated that 'the soldier is a peculiar animal that alone can be brought to the highest efficiency by inducing him to believe that he belongs to a regiment infinitely superior to the others around him'.[73] The stoicism of the 'British Tommy' has been remarkable, repeatedly redeeming the poor leadership of his officers and perennial Government failure to provide sufficient funding for the Armed Forces. Dixon argues that soldiers' fortitude frequently rescued senior commanders and politicians from disaster and disgrace.[74] In refusing to recognise 'shellshock', General Sir Hubert Gough appears to fulfil the stereotype of the uncaring, Blimpish First World War senior commander:

> It is inconceivable how men, who have pledged themselves to fight and uphold the honour of their country, degrade themselves in such a manner and show an utter want of spirit and courage which at least is expected of every soldier and Britisher.

Not all the rank-and-file were lions, nor every senior commander a donkey, a characterisation that, as Bond argues, has become received popular wisdom thanks to the 1960s lens through which the First World War has been viewed.[75] Lord Percy, Colonel of the 5th Foot, undoubtedly felt a sense of identity, loyalty and responsibility to his subordinates: in 1776, he gave the widows of soldiers killed at Bunker Hill $5 each for their passage home, with an additional five guineas to be distributed to each of them when they landed. He also spent £900 on one commission and 'purchased others for excellent men and good officers who must otherwise have their juniors raised over their heads'.[76] Percy reflects the strength of the paternalistic bond that could exist between a senior commander and his military 'family'.

The Bond of Loyalty

This section will explore the Military Covenant's claim that a bond of loyalty – steadfastness in allegiance – has existed throughout the Army's history between soldier, Army and nation. Mutiny and desertion are two metrics of a soldier's loyalty to the Army as an institution, while the Army's loyalty to the nation is demonstrated by its subordination to the State and respect for civil–military norms. To gauge the bond of loyalty between nation and soldier, for the purposes of this chapter, the absence of loyalty – a sense of betrayal – will be assessed. This will be explored in

72 French, 2004, citing letter to *The Times*, 1 December 1854, p. 10.
73 Wolseley, 1880, p. 2.
74 Dixon, 1976, p. 207.
75 Bond, 2002.
76 *St George's Gazette*, 1890, p. 41.

the following section in the context of whether the nation has discharged its moral responsibility to soldiers in exchange for their offer of sacrifice.

The English Civil War and Protectorate cast a long shadow over British civil–military relations, living on in folk memory and engendering an irrational fear of militarism for two centuries. The Bill of Rights following the Glorious Revolution of 1688 declared that a peacetime standing Army was only legal with the consent of Parliament. 'That single clause removed the standing Army from the Royal household and placed it firmly under the control of Parliament, making it a national institution'.[77] The Mutiny Act (1689) and the Army Act (1713), both of which had to be passed annually, reinforced the legislature's control. Despite this, the Army continued to be regarded as a potential instrument of tyranny, particularly by successive generations of the Commons' squirearchy, unplacated by Prime Minister Pelham's assertion in 1744 that: 'Our liberties are in no danger from our standing army because it is commanded by men of the best families and fortunes'.[78] As Spiers suggests, constitutional distrust of standing armies is a 'cornerstone of the English libertarian tradition'. Regarded as the first line of home defence, the Militia not only allowed local social elites to keep any armed forces raised under their control, but was also a counterbalance to the potential increase in State power represented by standing forces. Gibbon declared during the Seven Years War: 'A martial enthusiasm seemed to have pervaded the land, and a constitutional Army was formed under the nobility and gentry of England'.[79] A Militia captain, he soon had reservations that it all too rapidly emulated the regular Army ('the mercenaries'), declaring that the Crown 'had acquired a second Army more costly and less useful than the first'.[80]

The Army's potential for disloyalty to Crown and Parliament was entrenched in British psyche, not least because of the rule of the Major Generals (1655–57). Oak Apple Day, 29 May, commemorating the Restoration of Charles II, remained in the Book of Common Prayer and part of Anglican services for almost two centuries until 1858. Just as 5 November keeps alive a vestigial anti-Catholicism, celebrating the Restoration was a reminder of the Army's potential for disloyalty to contemporary constitutional arrangements. The construction of barracks during the 1790s caused national alarm: radical MP Francis Burdett complained that they enabled 'troops paid by the people to subdue the people'.[81] However, the alternative, billeting, had been one of the complaints set down in the 1628 Petition of Right, which condemned the 'great companies of soldiers and mariners' dispelled throughout the country 'against the laws and customs of this realm and to the great grievance and vexation of the people'.[82] Suspicions about the Army's

77 Spiers, 1980, p. 57.

78 Alan J. Guy, The Army of the Georges, in Chandler and Beckett (eds), 1994, p. 103.

79 Spiers, 1994, p. 122.

80 Gibbon, 1984 edition, p. 124.

81 Holmes, 2001, p. 51.

82 Barnett, 1970, p. 71.

potential for disloyalty appear, in retrospect, overblown and curious. As Strachan observes, throughout its history, the Army has been one of the 'best-represented occupational groups' in Parliament.[83] In the 1790–1820 cohort of MPs, 107 out of 400 were Foot Guards, which 'would suggest a close link between wealth, birth and military-cum-Parliamentary activity'.[84] This indicates that according to Finer's taxonomy of military incursion into politics, the Army was undoubtedly influential. However, there was never any suggestion after the Restoration that it came close to 'displacement', i.e. 'a failure to defend the civil authority against violence', let alone 'supplantment'.[85] In addition, with Purchase, the Army was perceived as being far more representative of the ruling class than the Navy, which demanded technical proficiency from its officers.[86] Despite this, doubts about the Army's loyalty lingered, certainly until the end of the Napoleonic Wars, although it was the Navy that mutinied in 1797. The debate about the strength of political motivation behind the Nore and Spithead mutinies – summed up as 'sedition or ship's biscuits?' – was echoed after the Invergordon mutiny of 1931.[87] However, the Navy was never perceived to offer the same level of potential political threat as the Army. Huntington suggests that it was not until the Cardwell reforms of 1871 that a reluctant Parliament finally overcame its doubts about the Army and conceded 'that Moltke might be a greater threat than Cromwell'.[88] As Colley observes, at George III's coronation in 1761, spectators had been offended by the presence of troops: by the end of the nineteenth century Queen Victoria 'incurred her subjects' disapproval if she was not accompanied on official occasions by military magnificence'.[89]

The social dislocation of the late eighteenth century tested soldiers' loyalty as the Army brought aid to civil power. With no Metropolitan Police force until 1829, soldiers had a constabulary role, including restoring order when the Riot Act was read and curtailing smuggling. Eighteenth- and early nineteenth-century soldiers faced similar legal challenges as their contemporaries today when involved in war among the people, with civil law always enjoying primacy. While the 1780 Gordon Riots quelled by soldiers were deemed a test of the mob's – not the military's – loyalty to government, half a century later, reformers considered the Army tainted in connection with the Peterloo Massacre. Far from being a relic of the pre-Victorian era when the Army was the primary force for maintaining law and order, a state of emergency was proclaimed six times between 1920

83 Strachan, 1997, p. 26.

84 Ibid., p. 27.

85 Finer (editor Stanley), 1992, p. 140.

86 Raban's *Passage to Juneau* describes how the plebeian Captain Vancouver asserted his authority over his aristocratic young midshipmen by setting them exercises in trigonometry.

87 Anthony Browne, Sedition or Ship's Biscuits?, *Mariner's Mirror*, cited by N.A.M. Rodger, 2005. Bell, 2005, pp. 75–92.

88 Huntington, 1957, p. 47.

89 Colley, 1984, pp. 94–129.

and 1970 under the Emergency Powers Act.[90] The reliance upon soldiers for the maintenance of civil order, especially during a time of social upheaval that occurred between 1770–1830, justified reformers' wariness of the Army as an agent of potential despotism. However, with measures such as the suspension of Habeas Corpus in 1794, the Government itself tilted towards tyranny. In addition, as Spiers suggests, the Army never aroused the political passions of abolitionism, franchise reform or the Corn Laws. After Waterloo, civil–military relations centred not on constitutional proprieties about the Army's loyalty, but grumbling about its bloated size and its drain on the public purse. In 1828, Alderman Waithman MP told the Commons that the cavalry's 8,000 horses were 'devouring the country'.

The bond of loyalty between Crown and Army is symbiotic, with the two institutions reinforcing one another. Both use manufactured ceremonies, rituals and traditions, often involving the other. While soldiers came to be perceived more favourably in Victorian era, the Queen did much to promote them in a positive light, describing herself as a 'soldier's child' and taking a lead in soldiers' welfare. Writing to War Secretary Panmure in 1855, she demanded more military hospitals:

> This is an absolute necessity and now is the moment to have them built, for no doubt there would be no difficulty in obtaining the money requisite for this purpose from the strong feeling now existing in the public mind for improvements of all kinds connected with the Army and the well-being and comfort of the soldier.[91]

Until the introduction of the Distinguished Conduct Medal and the Victoria Cross, gallantry medals were issued only to officers. The Victoria Cross was introduced by Royal Warrant: the Queen sidelined Parliament insisting, incorrectly, that it was 'a matter which clearly belongs solely and entirely to the discretion of the Crown'.[92] A soldier's loyalty is to his unit, his regiment and then the Army, but also to Queen (or King) and Country. The award of a Waterloo medal upset Peninsular veterans 'livid that their own long sufferings had not yet been recognized with any badge or distinction'.[93] National Servicemen who served in Egypt's Canal Zone were similarly aggrieved until they were finally awarded their medal in 2003.[94]

The test of a soldier's loyalty to the Army is the incidence of desertion and mutiny. Unrest, let alone mutiny, in the 350-year history of the Army are strikingly rare. The Curragh 'mutiny' of 1914, when Irish officers threaten to resign en masse rather than be sent to Northern Ireland during the attempted enforcement of Home Rule, is perhaps one of the most serious incidents. During the First World

90 Babington, 1990, p. 123.

91 Sarah A. Tooley, *Personal Life of Queen Victoria*, Hodder and Stoughton, 1896, cited by Smith, 2008, p. 13.

92 Letter to Panmure 14 January 1856, cited by Smith, 2008, p. 41.

93 Urban, 2003, p. 200.

94 Hickman, 2004, p. 203.

War, the poor standard of accommodation and facilities caused unrest in the so-called New Armies' training camps, with mass meetings and strikes at two bases, Seaford and Codford in late 1914.[95] However, in the context of a 14-fold increase in Army numbers, with 5.7 million passing through its ranks, these incidents are insignificant.[96] The so-called Salerno Mutiny of 1943 seems to have been prompted by loyalty between the chain of command, in this case members of the Highland Division and their commander. Their defence counsel stated:

> To these men ... General Wimberley is their military god and the sign HD is the altar at which they worship. So much so, that to ask a soldier of the Highland Division to fight with another division is, in my mind, akin to asking a Hindu to worship Mohammed.[97]

While mutiny has been rare, desertion has not. Abolished in 1930, the reinstatement of the death penalty for desertion was demanded by Generals Fraser and Auchinleck when the rate rose from 6,889 in 1940 to 20,248 the following year.[98] In 2007, the House of Commons heard that 11,000 soldiers had gone absent without leave since the start of the Iraq War, 'with 1,000 still unaccounted for',[99] which implies that desertion is not unknown in today's Army. Desertion reflects a breakage of the Covenant between soldier and the Army and the soldier and his or her comrades. Overall, however, the bonds of loyalty between soldier and Army have been stronger than the loyalty shown by the nation to either.

The Bond of Responsibility

Responsibility is the third strand of the bonds linking Nation, Army and individual soldier within the Military Covenant. This section will explore the issue in the sense of moral responsibility the three component agents have evinced for one another: primacy will be given to the 'duty of care' nation and Army have shown towards soldiers. Soldiers not only have to sacrifice some of their civil rights – as well risk making the 'ultimate sacrifice' – but in their Oath of Allegiance swear to 'observe and obey all orders of Her Majesty, her heirs and successors and of the generals and officers set over me'. Consequently, although individual soldiers remain moral agents, their autonomy is curtailed. This section explores the support given to soldiers in the context of pensions and healthcare.

---

95 Peter Simkins, The Four Armies 1914–1918, in Chandler and Beckett (eds), p. 252.

96 Ibid.

97 Strachan, 1997, p. 213.

98 Brereton, 1886, p. 172.

99 Nick Harvey MP, 26 June 2007, HC Col 1560.

Since the Covenant's migration, civilians have disregarded soldiers' reciprocal obligations to the nation and Army that the concept codifies. One area of that responsibility is compliance with international norms concerning armed conflict, whether the *jus in bello* tenets of the Just War tradition or, from the beginning of the twentieth century, the international laws and conventions regarding armed conflict. The loss of military discipline – most notoriously at Badajoz in 1812 – has been, like mutiny, rare. However, not all soldiers are 'verry parfait gentil' knights, reflected most recently by the Aitken and Gage reports.[100]

In emphasising the historical nature of the duty of care evinced by the nation towards soldiers, both the Military Covenant and its civilian advocates have invoked an idealised past that does not withstand scrutiny. Since the Covenant's migration, the Conservative Party, the Haddon-Cave Inquiry and the Royal British Legion have all cited Elizabethan legislation as evidence that, historically, the nation has discharged its moral obligation to soldiers, as the Covenant formalises.[101] The Act for the Necessary Relief of Soldiers [and Mariners] was passed in the 1593 session, but was superseded, first in 1597 and then in 1601. As Hudson observes: 'The [1593] Act provided for parochial, nationwide, compulsory rates to provide disabled soldiers with life-long pensions'. The Conservative Party implies that the Covenant is part of 'our historic tradition', stating that the 1601 Act is 'the equivalent of today's Military Covenant'.[102] It is neither: it is an example of the 'past being used in service of the present'.

The Elizabethan Acts should not be regarded as a precedent for the twenty-first-century Covenant. While the Elizabethan legislation appears to put this principle on a statutory basis, the Roman example of granting land as a military pension could equally be invoked as a precedent. Elizabethan England had no standing Army; instead expeditionary forces were recruited ad hoc by Crown prerogative, with a 'spike' after 1585 when English forces became actively engaged in Europe and Ireland. Most soldiers were impressed rather than volunteers. As Hudson suggests, the Privy Council had previously passed Orders to provide pensions for disabled veterans – including from funds raised by butchers' Lenten licenses – but these had proven to be inoperable.[103] Despite three Acts in eight years relating to disabled soldiers' pensions, the veterans themselves were at the mercy of an Elizabethan version of the postcode lottery. Back-dated to 1588, the 1593 Act actually reflected the lack of moral obligation the 'nation', that is, the civilian population, had felt towards disabled veterans after the 1588 Armada. Statute was the recourse of the Elizabethan Council and Parliament because moral suasion had failed. When it came to discharging its obligations to disabled veterans, civilian

100 Aitken Report, 2009; Gage Report, 2011 (Chapter 6).

101 Haddon-Cave, 2009.

102 Conservative Party, 2008, p. 11.

103 Geoffrey Lewis Hudson, 1998, pp. 87–93.

society had to be pressed as much as the soldiers who had served the nation. The 1593 Act was revised in 1597 and then in 1601 because it failed to raise sufficient funds through local parishes to provide for the disabled. The Elizabethan State's motives for introducing the Acts were primarily political: the internal and external defence of the realm, as the 1593 Act implies:

> … for as much as it is agreeable with Christian Charity Policy and the honor of our Nation, that such have since the 25th day of March 1588, adventured their lives and lost their limbs or disabled their bodies, or shall hereafter adventure their lives, lose their limbs or disable their Bodies in the Defence and Service of Her Majesty and the State, should at their return be relieved and rewarded to the end therefore that they may reap the Fruits of their good Deservings and others may be encouraged to perform the like Endeavours.

Shortly after the Armada, Admiral Lord Howard of Effingham stated that discharged sailors deserved to be 'better cared for than to starve and die'.[104] Pension provision was seen as necessary for preventing desertion, for keeping up the morale of those currently serving and to encourage others to serve. In addition, poverty was perceived as a threat to public order: the authorities feared rebellion among aggrieved veterans, especially during the economic dislocation of the 1590s.

Today's soldiers expecting the nation to fulfil its responsibility to them as formalised in the Military Covenant might be disheartened by the precedent that is claimed to have been set by the Elizabethan Acts. The Acts fell into abeyance during the Civil War, but were reverted to under William III, remaining on the Statute Book until 1758. Resurrected, they replaced 'the first systematic pension scheme in the British Army' introduced by James II, which built on measures introduced during the Commonwealth.[105] The Elizabethan Acts were a lottery: they relied on the authorities and apparatus in individual counties – JPs, County Treasurers and the Quarter Sessions – for their implementation. Some counties could not or would not comply with the Acts' provisions. In 1597 Norfolk refused to pay the pension of a volunteer; one former soldier was disowned first by Hertfordshire where he had lived, and then by Middlesex where he was impressed. Northumberland refused to collect the tax from the parishes.[106] In July 1632, JPs in Hampshire complained to the Privy Council that the 'number of maimed soldiers is so much increased since the actions of Cadiz and Re, that they have more pensioners than money to pay them'.[107] Some 'veterans' were found to be fake: a Westmoreland Justice informed the Privy Council that, 'Anthony Lucas recommended by them for relief as a maimed soldier is an idle, drunken peddler'.[108] Men such as William

104 Hudson, 1988, p. 107.
105 Brereton, 1886, p. 17.
106 Cruickshank, 1946, p. 129.
107 Secretary to the Commissioners of Royal Hospital Chelsea, 1872, p. 119.
108 Ibid., p. 117.

Wyatt mislaid their certificates proving they were soldiers. In addition, reluctant cathedral and college authorities had to be pressured into housing disabled veterans in their alms-rooms. If local authorities failed to implement the Acts, there was little that the veteran, or indeed, the Privy Council could do. Northumberland's non-payment went unchallenged, as Hudson surmises, because, as the first line of defence against Scotland, the county's support was needed.

Founded in 1681 by Charles II for the 'succour and relief of veterans broken by age or war', Chelsea's Royal Hospital appears to embody the fulfilment of the nation's moral responsibility towards soldiers who had fought on its behalf. Today, the Hospital's fundraising literature declares that it is 'Keeping the Nation's Covenant with Our Old Soldiers'.[109] For much of its history the Wren-designed Hospital actually represents the nation's absence of moral responsibility to aged, infirm and disabled former soldiers. Like the Elizabethan Acts, Chelsea's foundation owed as much to political expediency as philanthropy. In the light of his father's experience and his own 'travels', Charles II was hardly unaware of the importance of keeping soldiers' support: the Restoration government's first act was to pay New Model Army veterans.[110] The £20,000 that Charles had promised for the Hospital never materialised. With the Hospital filled to capacity when it opened, James II introduced the concept of the out-pensioner: a veteran whose living expenses would be met while he waited for a vacancy. Until the First World War, all Army pensions, such as they were, would be administered by a Board of Commissioners at Chelsea. The Hospital, which emulated Les Invalides in Paris, symbolised Charles's sense of moral responsibility to former soldiers but, politically more significant, his commitment to a standing Army. As Wren stated: 'Architecture has its political use'.[111]

Chelsea represents how soldiers have looked after their own.[112] The Hospital's official papers noted in 1847: 'The requisite funds for carrying out his [Wren's] designs were almost entirely provided by the officers and men of the Army itself'. Two years after the foundation stones for the Hospital and its sister institution at Kilmainham, Dublin, were laid, a 10 per cent tax on officers' commissions was levied for building costs, as well as a one-off payment of five per cent of all Army pay. The buildings' maintenance continued to be funded by deductions from Army pay until 1847.[113] The cost of out-pensions rose steadily; from £5,672, 1s, 2d in 1705 to £1.4m in 1828: by the end of the American War of Independence, there were more than 20,000 out-pensioners, outnumbering the

109 Royal Hospital Chelsea, 2009.

110 Wordern, 2009, pp. 145–165.

111 Parentalia 1750, cited Secretary to the Commissioners of Royal Hospital Chelsea, 1872, p. 44.

112 Paraphrasing General Sir Mike Jackson after the 2005 rescue of two SAS soldiers: The Army looks after its own. Underline that three times, Shipman, 2010.

113 Brereton, 1886, p. 17.

mainland peace establishment.[114] However, it was not until 1798 that the money from the Exchequer to pay for the pensioners exceeded the total contributed by the military. Between 1681–1730, Chelsea had received £1,369,110 2s 7d from the Forces, compared with £384,376, 15s 3d from the Exchequer. By 1765 this had risen to £3.9m from the Forces and a total of £646,557 1s 5d from the Exchequer. In 1798 the Forces' total contribution to Chelsea amounted to almost £5.7m, exceeded by the Government's £5.9m.[115] In addition, Chelsea – and its pensioners – should have been the beneficiaries of Prize money, but as was noted:

> A liberal and permanent endowment would have accrued from the unclaimed share of Prize money, the Statutory property of the Hospital, had not these funds been diverted in past years to other though kindred purposes by the financial exigencies of the State.[116]

Chaplain-General Gleig stated that the Royal Hospital belonged to soldiers: 'The establishment is his own, built by his own or his predecessors' money, supported out of funds which the nation never gave'.[117]

Parliamentary involvement in Greenwich and Chelsea legally formalised the nation's obligation towards its veterans, 'broken by age or war'. While both institutions ostensibly symbolise the nation's sense of responsibility, they could also represent the tacit need to keep veterans under some form of control. As Gerber argues, 'societies have long been haunted by fears of the violent potential of veterans with unpredictable mental states'.[118] As Waller suggests:

> The veteran is always a powerful political force, for good or evil, because others cannot protect themselves from him. He has fought for the flag and absorbed some of the *mana*. He is sacred. He is covered in pathos and immune from criticism.[119]

The magnificence of both buildings is a reflection of the State's power, while their charitable purpose would have deflected any criticism of it.

The sense of responsibility owed to soldiers – both serving and retired – has usually mirrored the ebb and flow of war and peace. Successive governments would revise pension arrangements during a time of conflict, only to attempt

114 Joanna Innes, The Domestic Face of the Fiscal-Military State, in Stone (ed.), 1994, p. 112.

115 Declared Accounts of the Chelsea Hospital, Secretary to the Commissioners of Royal Hospital Chelsea, 1872.

116 Ibid., p. 8.

117 Rev G.R. Gleig, ibid.

118 Gerber, 2000, p. 7.

119 W. Waller, *The Veteran Comes Back*, Dryden Press, New York cited by Jones and Wessely, 2005, p. 164.

to retreat from their commitments once a peace treaty was signed. In 1806 Wyndham's Act introduced a raft of standardised pensions, but officers rather than the rank-and-file were beneficiaries. Peninsular veteran George Baller of the 95th Rifles was not atypical; living in 'grinding poverty and great physical pain', with a pension too small to provide for his children.[120] An 1822 Parliamentary return highlights how rank on retirement determined the size of a pension. Lieutenants Hamilton and des Voeux both of the 20th Foot, both lost a leg in Holland in 1799: in 1811 Hamilton, then a Colonel received £300 a year, compared with Captain des Voeux's £100.[121] The document reflects that pension administration remained haphazard: Lt Barrow of the 4th Foot, wounded at Bunker Hill in 1775, had to wait until Christmas Day 1811 to receive his £70 annuity. In 1819 a Royal Proclamation ordering the pensions of hundreds of former soldiers to be cut, was condemned in the House of Commons two months later. Lord Nugent MP stated that veterans had such 'claims on the justice, the sympathy, the humanity, and the gratitude of this country, it would be difficult to conceive any of a more sacred or imperious character'. However, with pensions costing the nation £1.4m in 1828, Wyndham's Act was axed, despite a Select Committee warning that it would be 'an apparent breach of faith to men who were now in the service' and that the pensions were 'engagements little short of positive compacts'.[122] Wyndham's Act was passed at the height of the Napoleonic Wars: more than 20 years later, pensions were no longer a peacetime government's priority.

Until soldiers themselves could exert political pressure rather than just moral pressure, successive governments would inevitably attempt to evade their responsibilities once the danger of war had passed. In 1864, pensions were introduced for those soldiers who had completed 21 years of service, but disability pensions remained discretionary until the Second World War. Between 1861–98 only one soldier in three who left the Army received a pension:[123] Spiers suggests a quarter of Edwardian Army veterans became unemployed or unemployed vagrants.[124] The plight of many veterans was described in the 12th Foot's regimental gazette in 1864:

> Is it not sad to see constantly an old soldier being discharged without one half penny in his pocket, without a suit of clothes to wear, and who when he sees the money subscribed by his comrades, immediately spends it in the nearest Canteen and opens his first night free (as he calls it) either lying helplessly drunk in a drain or else in a neighbouring lock-up.[125]

---

120 Urban, 2003, p. 206.

121 War Office Return, 18 April 1822, National Army Museum.

122 Select Committee on Public Income and Expenditure, cited by Abbott Parry and Codrington, 1918, p. 27.

123 French, 2004, p. 252.

124 Spiers, 1980, p. 193.

125 *East Suffolk Gazette*, 1864, p. 3.

The help given to the penniless veteran by his comrades reflects their sense of moral responsibility towards him – and perhaps their rational self-interest – but also the absence of any support mechanism on the part of the nation.

The nation's support in exchange for military service, as codified by the Military Covenant, has never been unconditional or consistent. Despite veterans' issues affecting a far higher percentage of the population with the mass mobilisation of 1914–18, pension provision and healthcare, particularly for the disabled, was half-hearted. The Lloyd George Coalition established a Ministry for Pensions in 1916 and, in July 1917, the Pensions Appeal Tribunal. However the onus was on claimants to prove that their 'disability is due to military service or is aggravated by it'.[126] In 1943, the government reversed the burden of proof, resulting in 70 per cent of claimants actually receiving a pension, compared with 30 per cent previously.[127] The post-war recession, which made a mockery of the wartime Coalition's 'Homes for Heroes' pledge, led to unemployment reaching two million, a quarter of whom were ex-servicemen. The 1921 launch of the British Legion, along with the presentation of an 825,000-signature petition to Parliament by Brunel Cohen MP in 1925 in support of a better deal for former Service personnel and their families, reflected the strength of support for veterans' issues. George Bernard Shaw stated: 'The duty of the country is perfectly clear. These men were disabled in its service and should be supported by it unconditionally'.[128]

While successive governments since the sixteenth century tried to evade their moral responsibilities or were begrudging in their attitude to veterans, soldiers themselves often rallied round, setting up subscriptions and raising funds for wounded, aged and infirm comrades, widows and orphans. In 1866, a dinner with the Duke of Cambridge raised £103 for the Asylum for Soldiers' Widows:[129] any surplus from the Rifles' Crimea Memorial and Centenary Appeal went to the Rifleman's Aid Society.[130] The same sense of 'regimental family' was reflected when Foot Guards' officers raised £9,000 to fund married quarters for 54 soldiers in 1852.[131] If a soldier died, his dependents' rations were supposed to be stopped after four months, explaining why widows of the rank-and-file tended to re-marry quickly, usually to one of her dead husband's comrades. Wyndham's Act entitled officers' widows to receive pensions. Between 1816–17, 1,651 Army officers' widows received a total of £81,458 from the state: 94 'general officers'' widows received £120 a year, while the widow of single hospital mate received just £26.[132] Exceptional among the traditionally scant provision for orphans who would otherwise have had to rely on the workhouse and parish relief was the

126 Gerber, 2000, p. 82.
127 Brown, 1971, p. 106.
128 Ibid., p. 26.
129 *Household Brigade Journal*, 1866, p. 76.
130 *Rifle Brigade Chronicle*, 1899, p. 6.
131 de Watteville, 1954, p. 188.
132 War Office Return, 20 April 1818, National Army Museum.

Royal Military Asylum at Chelsea founded by the Duke of York in 1801, which became home to 1,500 orphans.

The public felt a growing sense of moral responsibility to soldiers and their dependents in parallel with its sense of identification with soldiers. From the latter half of the nineteenth century, this was expressed through charitable donations. The Royal Patriotic Fund, established during Crimea, rapidly raised more than £1.5m for soldiers' dependents. Apart from the short-lived Commonwealth Ordinances, the Fund represented 'the earliest national recognition of the claims of widows and orphans for war pensions'.[133] However, these Crimea widows – of whom there were still 1,314 in 1896 and who had to accept that 'no claim to relief is acknowledged as a right' – received just five shillings a week.[134] The rise of the popular press in the late Victorian era helped galvanise support for the charities, including the 1880s Zulu War Fund and the Imperial War Fund. Kipling's *Absent-Minded Beggar* written for the Daily Mail Fund was set to music by Arthur Sullivan and raised £250,000 for the gentlemen in khaki ordered South, 'doing his country's work' during the Boer War.[135] The lyrics stress the nation's obligation to soldiers, reminding the public that they must 'pay, pay, pay' for soldiers' welfare – and for the welfare of their dependents. Throughout the Army's history, the existence of charities connected with soldiers' and dependents' welfare points to successive governments' failure to fulfil their moral responsibility towards those who have fought for the nation. Registered in 1944, the Army Benevolent Fund was set up 'to give help where State assistance is either inapplicable or inadequate'.[136] The continuing generosity of the civilian public can be contrasted with the cheese-paring attitude of successive governments, reflecting the disparity of attitudes within 'the nation'. The reliance on far from guaranteed charitable donations is a reproach to the parsimony of the State rather than the people.

The healthcare given to soldiers and veterans is another metric of the sense of responsibility felt by the nation and the Army towards them. Naturally, medical care has to be judged by the often inadequate standards of the time under review. Yellow fever saw the West Indies regarded as the graveyard of the Army in the late 1790s, while between 1776 and 1780, more than 10 per cent of the troops sent there died at sea.[137] In the Walcheren campaign of 1809, there were 217 combat casualties but 23,000 deaths in four months from disease.[138] Cholera, dysentery and typhoid swept through the Crimea, with a disease rate of 161.3 per thousand for British troops, 119.3 for Russians and 253.5 for French forces.[139] Synonymous with the inadequacies of healthcare, Crimea saw the highest proportionate battle

133 Abbott Parry and Codrington, 1918, p. 59.
134 Ibid., p. 43.
135 Rudyard Kipling, *The Absent-Minded Beggar*, 1899.
136 Brereton, 1886, p. 185.
137 Neuberg, 1989, p. 108.
138 Lovegrove, 1952, p. 7.
139 Gabriel and Metz, 1992, p. 170.

losses and the highest disease loss rate than any other previous war for which there are records, with a French hospital mortality rate equivalent to that of the Middle Ages.[140]

Soldiers' healthcare in theatre was historically prompted often not by any great sense of compassion on the part of senior commanders, but by operational necessity. In a dispatch to London, Wellington stated:

> One of the reasons that induced me to cross the Tagus on 4th August instead of attacking Soult, was the want of surgeons with the Army, all being employed with the hospitals, and there being scarcely one for each brigade: if we had had an action, we should not have been able to dress our wounded.[141]

Although a chain of hospitals was established along the route to Salamanca, similar to the provision made during Marlborough's campaigns a century earlier, medical support was patchy. In the immediate aftermath of Waterloo, many of the wounded had to make their way on foot to Brussels to tents in the main square. According to the Deputy Inspector General of Hospitals George Guthrie: 'Nothing could recall the past irretrievable mischief that insufficient medical care had occasioned in the first few days'.[142] Although it was not until the Franco-Prussian War that there was a conflict of any magnitude in which more soldiers were lost to hostile fire than to disease,[143] it was only when ordinary troops began to be valued by the nation – and by indeed their senior commanders – that adequate medical provision for the Army became an imperative. Gabriel and Metz argue that medical care for the soldier became a two-fold necessity; to preserve the army's fighting ability in the field and to preserve the civilian population's support for military adventures.[144] The provision of healthcare in the Boer War, which included ambulance trains, reflected not just the medical advances since Crimea but the change in attitude towards the soldier who was fighting on behalf of the nation.

Conclusion

On 11 November 1920, Earl Haig stated: 'It is only by discharging fully our obligations to living ex-Servicemen, and to the dependents of all who fought for us, that we can hope to square our account with those who gave their lives for us and ours'. Throughout the Army's history, the nation has rarely squared its account with soldiers who have fought on its behalf. This lack of a sense of moral responsibility can in part be explained by the nation's lack of a sense of identity

140 Ibid., p. 170.
141 Martin Howard, 2002, p. 29.
142 Lovegrove, 1951, p. 8.
143 Gabriel and Metz, 1992, p. 147.
144 Ibid., p. 271.

with the rank and file until the late nineteenth century and by its longstanding suspicion about the Army's loyalty. However, the nation could have taken a lead from many senior commanders who showed little sense of identity with the men under their command. Perhaps commanders such as Marlborough, who understood the Military Covenant's 'common bond of responsibility' between themselves and their soldiers, are remembered because of their rarity. He stated that 'good order and military discipline are the chief essentials in an Army', but warned that they could only be preserved 'if its soldiers have meat in their bellies, coats on their backs and shoes on their feet. All these are as necessary as arms and munitions'.[145] Too often, commanders emulated Wellington, who regarded the Army as a 'huge and extravagant reformatory'.

The use of somewhat archaic language, along with the co-option of the past, undoubtedly lends the Military Covenant gravitas and authenticity, but its historical claims are open to challenge, especially in the light of the assertion that the bond of identity, loyalty and responsibility between Army, soldier and nation is 'unbreakable'. Although frequently tested in the Army's 350-year history, soldiers have generally remained loyal to the institution – although this is often expressed in the context of the regiment – while showing a sense of responsibility towards, identity with, and loyalty to their comrades. However, claims concerning the nation's 'unbreakable' bonds of responsibility, identity and loyalty to soldiers cannot be substantiated. In connection with the nation's regard for the welfare of soldiers, both serving and veteran, history is a poor blueprint for twenty-first-century public policy.

145 Brereton, 1886, p. 23.

Chapter 2
The Doctrinal Context

Introduction

Military doctrine provides a corporate conceptual framework for analysing the use of force. The formalisation of doctrine is a comparatively recent development in Britain, beginning in 1989. *Soldiering: The Military Covenant* was published in February 2000, the fifth and final volume in a series of Army Doctrine Publications, which had begun in 1995 with *Operations*.[1] Exploring the 'essential truths of being a soldier',[2] *Soldiering* was written by (then Brigadier) Sebastian Roberts, as a background paper to *Values and Standards of the British Army*.[3] Complete or abridged editions of this work, updated in 2008, have continued to be issued to all soldiers: consequently, within the military sphere, *Values and Standards* has always been far better known.

This chapter examines the genesis of the Military Covenant from the mid-1990s within the context of what military doctrine had identified as the 'Moral Component' of Fighting Power. It will explore how *Values and Standards* and *Soldiering*, the core publications of the Moral Component, represented an ethical stock-take by the Army. With all doctrine having a potential readership among civilians, any examination must be mindful of its institutional role in promoting and defending the Army, particularly to policy-makers. The Army's assertion of its right, then of its need, to be different from civilian society in the context of the 'legal component' will be assessed. Above all, the development of the Moral Component represents the Army setting out its institutional vision of the civil–military relationship.

Military Doctrine: Development

The formalisation of doctrine in 1989 is judged by some analysts to be a radical development. Sheffield suggests the 'reinvention' of the Army as a doctrinally-based organisation is 'as profound a revolution as any experienced' in its history.[4] In *British Military Doctrine*, the Army's first doctrine publication, CGS General Sir John Chapple acknowledged that 'some say that laying down doctrine like this

1 *Operations, Command, Logistics, Training, Soldiering.*

2 Major-General (Retired) Sir Sebastian Roberts, interview.

3 ADP, 2000b. Army Doctrine Publications. Ministry of Defence. © Crown copyright 2014.

4 Sheffield, 2005.

is not the British way'.[5] In *BMD*'s second edition, CGS Guthrie stated that the original had broken new ground: 'We had not before sought to articulate doctrine at any level above the tactical'.[6] *BMD* was commissioned by Field Marshal Sir Nigel Bagnall, 'one of the most significant reformers in the 300-year history of the British Army',[7] who also established the Army's Higher Command and Staff Course (HCSC). Former CGS, General Sir Roger Wheeler, suggests that the formalisation of doctrine started with Bagnall's arrival as a Divisional Commander in Germany in the late 1970s:

> It's said he asked for the Operational Defence Plan and the reasons behind it and was told, 'Well, that's how it's always been'. He said that's not a good reason and set about to developing a completely different doctrine ... He then put his ideas into practice in the great Corps exercise that year and demonstrated that the previous concept of operations was far too static.[8]

The author of the first *BMD*, (then Colonel) Timothy Granville-Chapman, who had been chosen for the task by Bagnall during the first HCSC, explains:

> The need for high-level doctrine became very evident in the first HCSC. There was much debate about it during the course. Its absence was notable. It was evident that we had simply not thought through why we were doing what we were doing in Germany and were unclear about a good deal at high level. So *BMD* was an entirely original piece of work and it is probably best known for the hierarchy of fighting power that was articulated in it. [The Physical, Conceptual and Moral Components] In fact, in its first chapter it also defined doctrine ('Doctrine is what is taught') and laid down for the first time that manoeuvre should be at the heart of the British approach to warfare. It emphasized Fuller's dictum on the need to develop people who knew *how to think rather than what to think.*[9] (italics, *sic*)

Major-General Andrew Sharpe, Director of the Development, Concepts and Doctrine Centre, suggests that Bagnall questioned the whole premise of the British Army of the Rhine and whether British Forces should be merely 'a speed bump for nuclear release'.[10] Since 1989, doctrine has become integral to the Army. *British Military Doctrine*'s central operational tenets – the Maneouvrist Approach and

---

5 ADP, 1989.

6 ADP, 1996, Foreword.

7 Sheffield, 2005.

8 Wheeler, interview.

9 Granville-Chapman, note to author (Major Granville-Chapman had been Bagnall's Military Assistant, when he had been the 1 British Corps Commander).

10 Sharpe, interview.

Mission Command – remain at the core of the latest statement of the Army's capstone doctrine, *Operations* (2010).

Traditionally, the Army relied primarily on informal rather than formal methods of disseminating knowledge. Best practice was passed on by 'regimental osmosis', backed up by 'official tactical manuals and some unofficial but influential writings'.[11] Today, this would be described as lower level doctrine, relating to tactics, techniques and procedures. Callwell's *Small Wars* acknowledged that 'upon the organization of armies for irregular warfare valuable information is to be found in many instructive military works, official and non-official'. One such work could have been Wolseley's *Pocket Field Manual*: while not addressing whether war is a continuation of politics with an admixture of other means, it included invaluable advice about how much forage camels and elephants needed each day.[12] Many contest that the Army was a completely doctrine-free zone before 1989. Sharpe suggests:

> The British Army wasn't good at publishing something called doctrine in a doctrine pamphlet … [but] Callwell was writing the stuff that would now be probably turned into doctrine.[13]

A retired senior commander observes: 'There were shelves of material. The reason why the myth has come about is that there was no clearly identifiable capstone doctrine'.[14]

The pre-1989 lacuna in higher-level doctrine has been attributed to two main factors: the British character and the varied nature of British soldiering, whether in Europe or the Empire. Sharpe suggests there was localised brigade-level doctrine, specifically written for one area, for example, Malaya. 'You wouldn't see that stuff being used in Kenya, you wouldn't see that stuff being used in Cyprus'.[15] Historically, British soldiers have been, in colloquial terms, doers rather than thinkers. Although the Royal Military Academy at Sandhurst can trace its origins back to the Staff College of 1858 and the Royal Military College of 1800, the Army has traditionally lacked a 'brain'; that is, institutions to analyse the concept of war. The Army shied away from theoretical study, with senior commanders wary that doctrine could become dogma. Instead, as Kiszely suggests, 'faith was placed in an empirical approach together with a belief that character rather than intellect was more important in war'.[16] Huntington observes that, traditionally, the ideal British officer was not unlike a *beau sabreur*, 'a man of boundless courage and audacity

11 Sheffield, 2005, p. 167.

12 Wolseley, 1886, p. 61.

13 Sharpe, interview.

14 Officer M, interview.

15 Sharpe, interview.

16 Holden Reid, 1995, p. 2-A-4; J.P. Kiszely, The Contribution to Military Success, in Holden Reid (ed.), 1993, p. 36.

but no reflection'.[17] French argues the unwillingness to be prescriptive was the result of cultural forces within society: 'To be British was to be pragmatic and to "muddle through"'.[18] Despite this, as Holden Reid observes, after 1918, Britain became a world leader in military thought, producing two military thinkers of 'first rank', Liddell Hart and Fuller;[19] the latter, 'could withstand comparison with Jomini and Clausewitz'.[20] If the Army had chosen to formalise a capstone doctrine, or indeed 'the British way in warfare', there were enough brains for it to pick.

The formalisation of doctrine after 1989 created a cultural shift within the Army, making conceptual thinking about their profession by soldiers acceptable, both institutionally and in the Mess. Lord Dannatt, formerly Colonel of the HCSC – itself a reflection of the new mood of intellectual inquiry within the Army – suggests:

> To admit to such an interest in your profession was considered a fairly appalling crime, and a somewhat ungentlemanly expression of trying too hard – we, perhaps, overdid the Corinthian spirit of the gifted amateur.[21]

Until the 1990s, graduates were in a minority at Sandhurst. Sharpe suggests that being bookish was looked down upon: in the Army he joined it was more important to be on the rugby field in any spare time. 'If you were studying the manuals you were a military shit'.[22] This has changed.

> I don't think young officers are now prepared to accept a General who stands up in front of them and says, 'I don't bother with doctrine, I just do it my way and my way is jolly good'.[23]

Today, doctrine and conceptual analysis are equated with professionalism.

The introduction of formal doctrine in 1989 is heralded as 'a new era' for the Army.[24] However, this is the judgement of hindsight: the development was overlooked by many contemporary commentators, including those writing in the *RUSI Journal*, who were, unsurprisingly, preoccupied by the end of the Cold War. One exception was the *British Army Review*, an 'in-house journal of professional thought',[25] which welcomed 'an authoritative statement of *what is required of the Army and how it will operate*' (italics, *sic*). It added:

17 Huntington, 1957, p. 53.
18 French, 2006, p. 50.
19 Holden Reid, 1998, p. 2.
20 Ibid., p. 63.
21 Dannatt, speech, 21 September 2007.
22 Sharpe, interview.
23 Ibid.
24 Sheffield, 2010.
25 Dandeker, observation.

> Some may say it tells them nothing new ... but it still needed saying and clothing in the mantle of authority to give form and direction to the development and progress of the study of our profession.[26]

In 1989, CGS Chapple emphasised that time was up for the traditional institutional disregard of analysis: 'A modern battlefield is not a place where we could hope to succeed by muddling through'.[27] In his 2010 foreword to *Operations*, CGS Sir David Richards stated:

> While this doctrine emphasises the importance of minimising prescription, the land operating environment is just too dangerous and complicated to make it up as we go along. I see the chain of command at both formation and regimental level playing a key role in educating subordinates in the importance and application of the key tenets of doctrine. This is a habit we must all acquire.[28]

Chapple's similar statement is a reminder that although doctrine might be codified at an institutional level, it is not necessarily inculcated or assimilated.

Although an exposition of Fighting Power, doctrine has also been a means by which the post-Cold War Army – mindful of a readership 'outside the wire' – could make its institutional case to civilian society. In 1989 Chapple stated:

> I hope, too, that it will be of interest to those outside the Army who need to know how we go about our business and seek to achieve success in today's conflicts.[29]

MacInnes suggests that 'the origins of British military doctrine were inherently political': conversely, the interviewees for this study suggest that they were grounded in the operational; i.e. confronting the Warsaw Pact forces in Germany.[30] Somewhat ironically, *British Military Doctrine*, focusing on the German plain and deterrence, was launched as the Berlin Wall started to crumble. However, doctrine also provided an additional means by which the Army could defend itself as an institution, especially to policy-makers, as well as to other states and their Armed Forces. Roberts says:

> Having lost the monolithic threat of Soviet Communism, the Army needed to be able to justify their existence and explain what they did: both to themselves and to outsiders from the public, the press, to our allies.[31]

---

26 Editorial, *British Army Review*, 1989.
27 *BMD*, 1989, Foreword.
28 ADP, 2010.
29 ADP, 1989, p. vii.
30 MacInnes, 2007, pp. 127–141.
31 Roberts, interview.

Gooch observes, that after 1918, there was 'no attempt to grapple with the greatest strategic lesson of the war' – that Britain's security was linked with that of Western Europe.[32] More than two decades after the end of the Cold War, despite the introduction of the *British Defence Doctrine*, which sits at the top of the doctrinal tree and 'sets out the philosophy of defence as an instrument of national power', the United Kingdom has been judged to lack an overarching national strategy.[33] Despite this, as MacInnes suggests, the development of individual Service doctrine in the 1990s provided, 'a ready-made, authoritative and approved justification for continued support of that Service'.[34]

The Moral Component – Development

'The purpose and measure of the British Army is military effectiveness: success in war and on other operations'.[35] This could be the Army's mission statement; a reminder that it is, above all, a fighting force. In 1989, *British Military Doctrine* identified the three inter-related Components of Fighting Power – 'an army's ability to fight' – as the Physical, the Conceptual and the Moral.[36] The Moral Component is 'the ability to get people to fight'.[37] Former Adjutant General, Sir Michael Rose notes:

> The willingness to fight is clearly made up of a complex series of elements. It comes from belief in a cause, loyalty to one's comrades, and an element of compulsion. It is founded on a distinctive military ethos, which has traditionally been based in the British Army on a complex system of education, training, and discipline.[38]

However, as *BMD* states, the moral factor is difficult to define: it involves not just morale, but motivation, leadership and management.[39] The two doctrine publications originally codifying the Moral Component, *Soldiering: The Military Covenant* and *Values and Standards of the British Army*, embrace the two definitions of 'moral': the concept of the 'good' in terms of ethical conduct and with moral in the sense of moral victory and moral support: 'perceptual or

32 John Gooch, A Particularly Anglo-Saxon Institution: The British General Staff in the Era of Two World Wars, in (eds) French and Holden Reid, 2002.

33 HC 435, 2010–11.

34 MacInnes, 2007, pp. 127–141.

35 ADP, 2000a, paragraph 1–1. Army Doctrine Publications. Ministry of Defence. © Crown copyright 2014.

36 *BMD*, 1989, p. 33.

37 Ibid.

38 Rose, email.

39 *BMD*, 1989, p. 35.

psychological rather than tangible or practical in nature or effect'.[40] Although the Moral Component has a practical, tangible end – Fighting Power – it addresses the intangible factors that contribute to it.

The codification of Moral Component was engendered by a number of issues, both internal and external to the Army, some of which reflected the existing tensions within the civil–military relationship. The institutional exploration of the Moral Component began in the mid-1990s, an era of downsizing and out-sourcing to civilian contractors. In comparison with the tempo after the start of Operation Telic in March 2003, the pressures on manpower in the post-Cold War and post-Gulf War Army of the 1990s were comparatively light, despite operations in Bosnia, Kosovo and Northern Ireland. Major-General (retired) Peter Currie suggests there was concern among senior commanders that the Army could become little more than a gendarmerie, a fate befalling Forces in Europe, notably in Holland.[41] Major-General (retired) Tim Cross, who had built refugee camps in Kosovo among other humanitarian operations in which he was involved, says:

> The primary purpose of the military is to win on the battlefield and we had to work hard at saying why. There are a number of threads running through all this but other militaries didn't stay at the high end of the spectrum.[42]

In 2000, the Chief of the Defence Staff, Charles Guthrie, stated the 'touchstone' of the Armed Forces was combat effectiveness.[43] However, for most of the 1990s, notwithstanding operations in Sierra Leone and Kosovo, any institutional justification couched in terms of combat effectiveness and battlefields could have sounded close to special pleading to many civilians. However, the Moral Component underscored the Army's institutional ethos:

> ... That spirit which inspires soldiers to fight. It derives from and depends upon high degrees of commitment, self-sacrifice and mutual trust which together are essential for the maintenance of morale.[44]

Both *Soldiering* and *Values and Standards* emphasise 'success in war and on other operations'.

Military doctrine was formalised, in part, to make the Army's case to the civilian sphere. Roberts says: 'The Army was addressing two audiences with the development of the Moral Component, soldiers and its "paymasters" in civil society'.[45] Currie explains: 'We needed a code for ourselves and a statement for

40 Merriam-Webster definition.
41 Currie, interview.
42 Cross, interview.
43 Guthrie, speech, RUSI, 2000.
44 ADP, 2000a, paragraph 301.
45 Roberts, interview.

everyone else about the British Army'.[46] The 1990s Army, like its sister Services, had to justify its existence – and its level of public funding. With the 1997 election of a Labour government, senior commanders were aware of paymasters to whom a war-fighting ethos was alien.[47] Before 'Blair's Wars', some Labour MPs were not necessarily the Army's natural allies. They were more likely to be peaceniks than their Tory counterparts, whether previous supporters of CND, Troops Out and the Greenham women. Although one minister was taken to task by the media for dismissing members of the Household Division – somewhat unimaginatively – as 'chinless wonders', his description of Trooping the Colour as 'doing incomprehensible things with flags' reveals the ministerial mindset.[48] Foreign Secretary Robin Cook told CGS Wheeler that the Army should be trained for peacekeeping rather than war-fighting.[49] According to Lord Guthrie, some ministers would become more supportive of the Army after the successful intervention in Sierra Leone and its post-conflict operation in Kosovo.[50]

The Moral Component was engendered by, and attempted to justify, the cultural and legal gap between the civilian and military spheres. With the end of 50 years of Cold War, the Army's *raison d'etre* was becoming less obvious to civilians, who were happy to take a financial peace dividend. In addition, social attitudes that would have been counter-cultural in the 1960s became mainstream. The fictional Bridget Jones captured the *zeitgeist*, arguing that the Labour Party best represented it:

> Labour stands for the principle of sharing, kindness, gays, single mothers and Nelson Mandela, as opposed to braying bossy men having affairs with everyone shag shag shag left right and centre and going to the Ritz in Paris then telling all the presenters off on the Today programme.[51]

Labour's first election landslide was emblematic of a Britain that was socially liberal, but correspondingly intolerant of sexism, racism and homophobia. Mileham, Deakin and Beevor underscore how the Army, an institution closely associated with traditional values, was out of step with the nation.[52] The Army appeared to represent the other, reactionary, side of the cultural divide from where the forces of Cool Britannia had amassed.

---

46 Currie, interview.

47 Roberts, interview.

48 Peter Mandelson, interview Marian Finucane, RTE, March 2000.

49 Wheeler, interview.

50 Guthrie, interview.

51 *Bridget Jones's Diary*, Fielding, 1996. The 'Bridget Jones test' was cited by David Willetts MP in 2010 as a metric of perceptions about the Conservatives.

52 Stephen Deakin, Education in an Ethos at the Royal Military Academy, Sandhurst, in (eds) Robinson et al., 2008, p. 19; Mileham in ibid.; Beevor, 1990.

The codification of the Moral Component grew out of studies undertaken by the Army from the mid-1990s concerning its right or need to be different from civilian society, necessitated by the demands of military operations. However, in attempting to convey the functional necessity of difference, the Army was in danger of alienating itself from the society it was supposed to serve and reflect. The Moral Component was initially explored during seminars convened by the Adjutant General, Michael Rose in 1995.[53] He says:

> What emerged was that the military ethos needed to be different from that of civilians, and that soldiers needed to be treated differently in law from civilians both in legal and ethical terms.

The resulting paper – *The Extent to which the Army has the Right to be Different*[54] – was endorsed by the Army Board the following year. However, while many officers shared Rose's view that the military ethos was 'under threat', some did not welcome the Army's claims to legal and cultural exceptionalism. Major-General (retired) Andrew Ritchie observes: 'It was overly assertive in the context of what was going on in Britain in the late 1990s'.[55] For some civilians, in claiming a right to be different, the Army was defending its right to be racist, sexist and homophobic. In 1993, the *Discipline and Standards Paper: The Military Ethos (The Maintenance of Standards)* had reaffirmed the ban on homosexuality; two years later the Commission for Racial Equality's investigation into the Household Cavalry was described as the 'nadir for race relations in the Armed Forces'.[56] Attempting to secure a Commons' debate on the racism issue, Diane Abbott MP cited the difference between the United States, where Colin Powell was Chairman of the Joint Chiefs and Britain, where, at the start of the decade, a Minister had given assurances that the problem was being addressed:

> Tell that to Jacob Malcolm, who in 1991 was barred from the Household Cavalry because of his colour … Tell it to Mark Campbell, who was the first black man in the Guards. In 1994, he was driven out by the taunts of 'nigger', the abuse, the violence and the bed soaked in human urine.[57]

Labour's commitment to equal opportunities in the Armed Forces was reflected in October 1997 by the Defence Secretary's announcement of an increase in the number of Army posts open to women:[58] the same month CGS Sir Roger Wheeler

53 Chapter 3.
54 Still classified.
55 Ritchie, interview.
56 HC HC29-I (2000–01) Paragraph 35.
57 HC Deb, 22 Jan 1998, c1247.
58 HC Deb 9 March 1998 c38W.

made a public commitment to end racial discrimination and re-introduce an equal opportunities strategy.

The Moral Component was an attempt by the Army to assert its professional autonomy, eroded by civilian incursions into the military sphere. Civilian law had penetrated the military's 'professional space',[59] a process examined by Rubin and Forster.[60] The incorporation of the European Convention of Human Rights into British law, with its impact on employment and equal opportunities practice, had 'massive implications for the disciplinary system', according to the co-author of the Service Test, Major-General (Retired) Stephen Andrews.[61] Ritchie suggests there was a need to 'educate ministers' to prevent the Army being 'suffocated' by legislation, including the Working Time directive. The Moral Component tried to convey its war-fighting ethos. Ritchie says:

> This was sent to ministers, they were asked to read it and understand it and discuss it and debate it. It was seen as a really important part of this wider battle of do you really want the British Army to become like some European Army where they're really suffocated and legislated against, which doesn't allow them to do anything.[62]

Both *Soldiering* and *Values and Standards* attempted to explain why the Armed Forces as a whole, but the Army in particular, should be considered a special case. As *Soldiering* states:

> There is a need to balance the demands of operational effectiveness and the ethos which underpins it, with the rights of the individual enshrined in legislation. The application of any legislation to the Armed Forces must be assessed in terms of its impact on the Moral Component of Fighting Power, so that appropriate exemptions can be sought where necessary.[63]

In 2000, CDS Guthrie warned against the Armed Forces being subject to 'inappropriate legislation'.[64] Andrews suggests that, within the Armed Forces, he observed 'a widespread reluctance to get to grips with the Human Rights Act'. However, the 'legal component' reflected changing attitudes in civilian society. Currie observes that there was also a strong mood within Labour to make sure the Army complied with legislation.[65] Ultimately, the European Court of Human Rights and employment tribunals to which, following the Wolfe reforms, Armed

59 A location explored by Burk and Forster.
60 Rubin, 2002; Forster, 2006 and 2012b.
61 Major-General Stephen Andrews, interview.
62 Ritchie, interview.
63 ADP, 2000a 2–1.
64 CDS Annual Lecture, RUSI.
65 Currie, interview.

Forces personnel could appeal, ensured the Forces' compliance.[66] The Army's battle against the legal component was a losing one.

Military Doctrine attempted to justify the seeming irreconcilability of the Army's ethos with the values of modern Britain. Underlining the competing pressures on the Army, former CGS Wheeler explains:

> It was absolutely essential we didn't build ourselves into a little isolated chunk of the UK, which was not representative of that society ... It doesn't necessarily mean you have to reflect all the standards of that society ... But whilst we want to be representative and understand its values, it was important to understand for operational effectiveness that there are times when the individual could not be pre-eminent.[67]

However, in attempting to identify, justify and assert difference, the Army risked denigrating civilian society. From the 1993 *Discipline and Standards* paper, cited by Deakin, it can be inferred that an 'us and them' attitude existed; 'we' were better than civilian society which had reduced in importance 'those values which the Armed Forces seek to maintain and regard so highly: sense of duty, loyalty, self discipline, self sacrifice, respect and concern for other'.[68] Analysts have observed a similar sense of military moral superiority developing recently in the United States. Bacevich suggests: 'Many in the military no longer care to protect our way of life – because they regard civilian life in America as degenerate and corrupt'.[69] *Values and Standards* says that the Army was recruiting from a 'less cohesive society' and 'one in which traditional shared values are less effectively transmitted and concepts such as honour and loyalty are less well understood'. *Soldiering* states 'the societies from which the British Army recruits have increasingly diverse ethical and moral codes'; previously these societies shared 'broadly common roots and horizons based on traditional, usually Christian, ethics and morals'.[70] The Army appeared ill-at-ease with civilian society, a discomfort doctrine has recorded.

Recruits represented the 1990s Britain by which the Army appeared institutionally disconcerted. The attitudes of rank-and-file recruits appeared to perturb their senior commanders as much as the malnourished Boer War volunteers had dismayed theirs. Sandhurst could no longer expect its cadets to arrive with

66 ECHR judgments which found the Armed Forces in breach of Article 6 included *Smith and Grady v UK*, 2000, *Lustig-Prean and Beckett v UK*. The latter case found the treatment of homosexuals in the Armed Forces, rather than the ban on homosexuality, was in breach of the Article.

67 Wheeler, interview.

68 Deakin, in (eds) Robinson et al., 2008, p. 20.

69 Bacevich, 2005, p. 23.

70 ADP, 2000a, paragraphs 102, 207.

ethical standards imbued from their family, public school education and faith.[71] In addition, the previous reliance on 'an enormous amount of informal, indirect education and training' and 'subliminal learning' to inculcate military ethics was no longer enough.[72] Potential officers – including women – come from far more socially and educationally diverse backgrounds than, for example, the majority of this study's interviewees. They are also more questioning of authority. Ritchie, a former Commandant of Sandhurst says: '"Ours not to reason why" no longer exists'.[73] In addition, the Army had become less of a vocation and more a career: there are few, if any, Sergeant Luke O'Connors to be found in the modern British Army.[74] Wheeler says he was concerned not only that some recruits had never played team games, nor understood teamwork and lacked physical fitness, but also that they 'came from a society in which the individual and individual rights were all-important'. The Moral Component provided the conceptual framework to explain to soldiers that they had to subordinate or forego some of their individual rights for the sake of operational effectiveness.

In addition to the external factors that prompted the exploration of the Moral Component, was the need for an ethical stock-take within the Army itself. An institutional malaise had led to a blind eye being too often turned to poor conduct within the chain of command. Andrews suggests that the Gulf War had, in part, provided the impetus for some institutional self-examination:

> We seemed to have two Armies, Northern Ireland and BAOR. They didn't meet very often. Too many soldiers were too fat to fight. Granby changed the culture. There had been no awareness of values-led leadership. We had to put across why we needed to have different standards from those in civilian life.[75]

Ritchie had a 'little black book' listing potential scandals involving senior commanders, which would have undermined the Army had they appeared in the media.[76] With more women becoming soldiers, a spate of liaisons took place between senior male NCOs and junior female officers. At the end of a trial that had lasted eight months, in 1996 three soldiers based in Cyprus were found guilty of killing a Danish woman. Stories were current about initiation ceremonies and bullying in Germany: a case of alleged rape at Shrivenham in 1996 and the Court Martial of a Lieutenant Colonel and a WRNS officer in 1998 provided yet more lurid headlines. In addition, the Army had resisted the introduction of compulsory drug testing. Andrews observes: 'As an institution the Army had not woken up to

71 Deakin, in (eds) Robinson et al., p. 17.

72 Ibid.

73 Ritchie, interview.

74 Ibid. Luke O'Connor VC, Alma veteran, was said to be married to the Army.

75 Andrews, interview.

76 Ritchie, interview.

the fact there was an issue with the standard of conduct of its people'.[77] Ritchie explains that while on an overseas tour, he had seen the malign effect on a unit when a sergeant was known to be having an affair with one of his subordinate's wives.

> There was a view – why is adultery and screwing around such a big deal? Look at Napoleon. People don't understand why this is so corrosive. It's about trust. There wasn't any framework around which people could talk about this and its importance to operational effectiveness.[78]

Ritchie says what he terms the ethical 'up-armouring' of soldiers, had come from CGS Wheeler. 'He devoted a lot of Army Board's time to this. It came from the top of the Army. Usually the focus is on the latest tank'.[79] Ritchie suggests the Army's values had to be set out:

> We had to make clear that it wasn't enough simply to shoot straight and to be fit. We needed to set out for the first time the rules of the road in terms of the personal values and standards of behaviour we required. And that you don't just spout them, you buy into them. If soldiers of any rank departed from the path, no-one could say they didn't know what they were doing.[80]

For most of the 1990s, claims to moral superiority by the institution could have been challenged by civilian society but were not. While many existing senior commanders were criticising the motes in recruits' eyes, they should have been removing the beam from the institution's. In addition, an implied claim to moral superiority creates the expectation that high standards will prevail: the view from the moral high ground should reveal how painful any fall might be. Andrews says: 'Deepcut told us that this had not permeated. There was still a lot of "do as I say, not as I do"'.[81] Many analysts have overlooked how the Army as a 'moral community' could be found wanting, engendering an examination of the 'moral' (i.e. ethical) aspect of the Moral Component.

Both *Values and Standards* and *Soldiering* represent the conflicting social pressures upon the Army between the end of the Cold War and the Iraq intervention, but also an institutional retreat – and subsequent counter-attack. Warnings by the Army throughout the 1990s that its ethos was imperiled by civilian laws, practices or values were ignored or overridden by governments, whether Conservative or Labour. Although he had stated in *Soldiering* that the Army should be exempt from certain legislation, Roberts concedes: 'It was becoming clearer and clearer

77 Andrews, interview.
78 Ritchie, interview.
79 Ibid.
80 Ibid.
81 Andrews, interview.

that special pleading was indeed difficult'.[82] Currie concurs: 'People in the Army recognized that change was afoot and also realized the importance of never finding ourselves isolated from society'.[83] However, paragraph 103 of *Soldiering* that sets out the Military Covenant, codifies not only that soldiers who forego their civil rights and offer sacrifice are different from those 'outside the Armed Forces', but that the Army 'differs from all other institutions'. The Covenant was, in part, a reassertion by the Army of its institutional claims to difference.

The Moral Component

Values and Standards

The first edition of *Values and Standards of the British Army* was produced in two formats: a 22-page version for officers (the focus of this section) that became the 'ethical bible' at Sandhurst and a four-page leaflet for all ranks.[84] Dandeker argues that the Army tried to develop 'a set of principles that would underpin policy responses to a range of people issues', rather than muddling through, 'responding to individual issues in an ad hoc way'.[85] The small group of senior officers who codified the Moral Component included Sebastian Roberts, author of *Soldiering: The Military Covenant*; Peter Currie who originally wrote *Values and Standards* and Andrew Ritchie who inherited Currie's draft: Army chaplain Iain Torrance was also involved. The draft form of *Values and Standards* predated *Soldiering*; some passages are almost identical, reflecting the collaborative nature of the development of doctrine. Although the two documents overlap, *Values and Standards* was, and remains, of greater magnitude to the Army than *Soldiering*. Significantly, it describes the Covenant as a 'two-way obligation' – between the Army and soldiers.

> Both share a common bond of identity, loyalty and responsibility for each other which is unwritten but unbreakable, and which has sustained the Army throughout its history.[86]

The exclusion of 'the nation' implies that the Army perceived the Covenant as being primarily an institutional, internally-focused concept.

Values and Standards underlines the unique nature of military service and the almost superhuman qualities expected of soldiers. Of the six core values that soldiers are expected to evince, CGS Wheeler chose to put selfless commitment

---

82 Roberts, interview.
83 Currie, interview.
84 Deakin, in (eds) Robinson et al., 2008, p. 23.
85 Dandeker, 2001.
86 ADP, 2000b, paragraph 11.

at the top of the list, followed by courage and then discipline.[87] Each value is described not just in the context of the individual soldier, but its importance is stressed in relation to being a member of a team:

> Like loyalty, respect for others goes both up and down the chain of command as well as sideways among peers. But it also extends to the treatment of all human beings, especially the victims of conflict, the dead, the wounded and refugees.[88]

Throughout the paper, onus is placed on the character and conduct of those up the chain of command to lead by example. However, in the context of integrity, the necessity of mutual trust between comrades is emphasised:

> They share the closest of quarters with their comrades; in tanks and in trenches they share food, drink and plastic bags for lavatories; they give each other blood and the kiss of life. They must have the profoundest respect for the individual not because of political correctness, but in the service of the greater good, because their cause and their lives depend upon it.[89]

The difference from civilian society is underlined in the emphasis on the functional necessity of respecting others.

Values concern individual character while Standards sets out a code for personal and interpersonal conduct to which soldiers are expected to adhere. The Standards section stresses the Army is different from most other employers, taking a different and more proscriptive approach to behaviour that would be regarded 'purely as a matter of individual choice or morality, and of no concern to the wider community'.[90] Proscribed conduct includes discrimination and harassment, bullying, drug misuse, alcohol abuse and irresponsible indebtedness, which reflects a 'lack of judgment and self-discipline'.[91] In addition, contact with the media, to 'express views on official matters or experiences' without the prior MoD approval is not permitted: 'Our responsibility to the Crown through Parliament means that the Army must be apolitical, and seen to be apolitical'.[92]

The abridged edition of *Values and Standards* distills in simpler, pithier language the 'ethos of the British Army'. Expected to abide by civil law, military law and war law, soldiers should also evince the six core values; selfless commitment, respect for others, loyalty, integrity, discipline, courage.[93] The individual soldier is depended upon by comrades, commanders and ultimately by the nation:

87 Wheeler, interview.
88 ADP, 2000b, paragraph 18.
89 Ibid., paragraph 16.
90 Ibid., paragraph 20.
91 Ibid., paragraph 28.
92 Ibid., paragraph 29.
93 Remembered as SOLID C.

'in short, they need to trust you and you need to trust them'. Significantly, the Military Covenant is described as a 'two-way obligation' – once again between the Army and soldiers. *Values and Standards* urges soldiers to reflect on whether their conduct would pass the Service Test. The Test itself, written by (then) Colonel Stephen Andrews and (then) Group Captain F.R. Amroliwala RAF, was introduced as part of the January 2000 Armed Forces Code of Social Conduct. It stated: 'Have the actions or behaviour of an individual adversely impacted or are they likely to impact on the efficiency or operational effectiveness of the Service?' It was deemed by many to be a more realistic 'take' on conduct than *Values and Standards* or *Soldiering*. 'It took all the bluster and sentiment out of the issue'.[94]

Both editions of *Values and Standards* ask a great deal from the individual soldier, with the acknowledgement that the Army is a far more intrusive and demanding employer than most others found in civilian life. While civilian employers can reasonably expect their workers to show integrity and respect for others, few if any, apart from a fire service, demand physical courage let alone selfless commitment. The values, or indeed virtues, expected of the individual soldier are somewhat idealised, with no consideration given to what might happen if a soldier falls short of the expected standards. In addition, with so much being asked, much is also promised to soldiers, as the paper states:

> In return you can at all times expect fair treatment, to be valued and respected as an individual, and to be rewarded by reasonable terms and conditions of service.[95]

However, neither document addresses the consequences of the bargain being unfulfilled, which came to be described by civilian society in terms of a 'broken' Military Covenant.

Soldiering: The Military Covenant

The core document of the Moral Component is *Values and Standards* for which *Soldiering: The Military Covenant* provided the 'philosophical underpinning'.[96] According to Roberts, he wanted to explore the essential truths about being a soldier, the most fundamental of which is the individual's responsibility as a weapons bearer who has to make decisions about the use of lethal force. As *Soldiering* states: 'All British soldiers share the legal right and duty to fight and if necessary, kill, according to their orders, and an unlimited liability to give their lives in doing so'.[97] Not only does such responsibility set soldiers apart from civilians, but, according to Roberts, it also sets them apart from personnel from the other Services. *Soldiering*, his own title for the paper, was chosen to maintain

94 Andrews, interview.

95 ADP, 2000c, p. 1.

96 Roberts, interview.

97 ADP, 2000a, Introduction.

this distinction between soldiers and others, despite the growing trend towards 'jointery'. Roberts says:

> All soldiers have to be prepared to give and take life, themselves, without somebody else saying 'yes' or 'no'. The document did always recognize the acutely personal nature of land warfare for every individual and that is its fundamental truth. In that sense it was designed very definitely to be proof against changing fashion or reality in warfare. …
>
> I remain personally convinced that there is a hugely important distinction between the realities of the environment for fighting, between land sea and air, which is so significant and so large, it needs to be very clearly defined. … We couldn't begin to produce anything valuable unless, above all, it was for us. Not for other armies, but for the British Army and not for other British Armed Forces. Not even for the Royal Marines, who could not be closer. We had to get it absolutely right for ourselves and then others could build on it or next to it, the things that were right for them. It was too important to be watered down for essentially political or economic reasons into joint doctrine. …
>
> To write truths about what it is to be a soldier has to be different from what it is to be in the Royal Navy or Royal Air Force or more particularly the truths about what it is to be in the British Armed Services. Which would become a pretty meaningless and amorphous document, not one which would tell you about why you die.

Soldiering was written by a soldier for soldiers. From 2007, civilians, rather than the Army or the other Services, championed the Covenant as a tri-Service concept. Generally, those civilians were unaware of the Covenant's provenance.

Soldiering was written in the context of an Army divided between Germany and Northern Ireland that had been part of the UN/NATO force in the former Yugoslavia. While there was consultation with others, not least because of the crossover between with *Values and Standards*, *Soldiering* was Roberts' own work. An Oxford history graduate and a Roman Catholic, described as 'the most cerebral soldier of his generation',[98] somewhat surprisingly, Roberts was unfamiliar with Clausewitz. He kept the Declaration of Independence ('some of the finest English ever written') and the Rule of St Benedict ('a set of standing orders, which have not changed for 1600 years') to hand while he was writing *Soldiering*. 'It deserved to be well written. If it hit the essentials of British soldiering, it should last for as long as British soldiering does'.[99]

98 Former officer.
99 Roberts, interview.

The Military Covenant affirms the Army's difference from civilian society and codifies the civil–military gap. However, *Soldiering* tries to explain the military to civilians. Roberts says:

> Without doubt, there are categorical differences which this document was designed to explain. ... It was definitely designed to be read by people outside the Army, particularly, politicians, civil servants, the media and the interested public.[100]

It also formalises the Army's expectation of the nation's material and moral support for soldiers, who should expect to be 'valued and respected' and 'rewarded by commensurate terms and conditions of service'.[101] According to the Covenant, there is no incompatibility between the military's institutional difference but receiving civilian society's support: consequently, it presents a paradox in the context of current gap analysis, much of which assumes that a divide between the civilian and military spheres is problematic. However, while affirming the Army's professional ethos, *Soldiering* was written as the Army faced external pressure to confirm to the legislation emanating from, and values reflected by, the Blair government. Any aspect of the gap that becomes unacceptable to the civilian sphere will be closed, reinforcing the norm of civilian control.

Soldiering analyses the Army's institutional ethos, judged distinct from that of the other Services. According to Roberts, 'we risk an awful lot by pretending our ethos is fundamentally the same'. It never loses sight of the Army's objective; 'fighting in the all-arms battle and achieving success in war'.[102] Consequently, the Army must remain structured and trained for 'full orchestra war' rather than becoming a gendarmerie.[103] Changes to the Army's ethos or to the 'terms, conditions or codes of service', will affect military effectiveness and the attitude of soldiers. 'Changing either will affect the output – the Fighting Power – which the Nation wants'.[104] Roberts emphasises that the uniqueness of soldiering means that the Army has to continue to press its case with government. He suggests: 'Because of the terrible nature of the responsibility, you need more urgently than ever to explain that there are places where only soldiers must go'.[105]

Soldiering examines the general moral principles relating to being a soldier. While touching on the justification for, and conduct of war, it is however primarily concerned with the individual soldier's conduct in the context of the institution – and on operations. With its echoes of Clausewitz – 'warfare and other military operations will remain essentially a battle of wills' – and aphorisms concerning

100 Ibid.

101 ADP, 2000a.

102 Ibid., paragraph 209.

103 Roberts, interview.

104 ADP, 2000a, paragraph 314.

105 Roberts, interview.

human virtue – 'you can't be good without being brave' – *Soldiering* aspires to be the 'higher level philosophical form of doctrine' identified by Palazzo.[106] The focus on moral dominance that leads to enduring success and lasting peace suggests a mindfulness of today's campaigning among the people in which British soldiers are involved. Roberts argues that all soldiers 'must always operate within and under the law, and whatever laws are laid on us, by, ultimately, the paymasters, our nation'.[107] In *Soldiering*, moral strength is also linked with just cause and international law, to which soldiers are subject:

> The duty of bearing arms, of being prepared to fight, kill and if necessary die in carrying out orders carries with it the responsibility to do so in a just cause … All British soldiers must discharge their duties not just according to orders and law, but consciously and clearly for the greater good.[108]

Soldiering anticipates one of the most problematic aspects of the Iraq intervention – the question mark over its legality, legitimacy and worth. Roberts suggests that the sacrifice soldiers are prepared to make needs to be perceived by civilians as justified. He says:

> It does remain true that if you have to be prepared to give your life and to take others on demand, you need to be able to explain that to yourself, your family, your spouse, as well as reporters and the public.[109]

Soldiering demands material and moral support for soldiers, yet again anticipating the difficulties that only a few years ahead would strain the civil–military relationship. It states: 'British soldiers must know that what they are called upon to do is right as well as militarily achievable, and has the support of the nation, society and the government'.[110] The Military Covenant migrated to the civilian sphere and was described as fractured because the tenets stated in *Soldiering* were disregarded in connection with Iraq and Afghanistan.

Although *Soldiering* sees trilateral reciprocity between Army, soldier and nation, it is corps and regiments that are seen as a 'key element' in delivering the Nation's and the Army's responsibilities.[111] The sacrifice expected of soldiers is emphasised, including the subordination of their individual aspirations, rights and needs to those of the team and the higher purpose. No matter their rank, they must be able to cope with 'ambiguity, uncertainty and change',[112] and they differ from

106 Palazzo, 2008, p. 12.
107 Roberts, interview.
108 ADP, 2000a.
109 Roberts, interview.
110 ADP, 2000a, paragraph 102.
111 Ibid., paragraph 315.
112 Ibid., paragraph 207.

civilian employees because the price of 'failure may be death'.[113] Echoing Hackett, the notion of 'unlimited liability' is stressed. In return:

> The Nation and the Army must fulfil their responsibilities in the Military Covenant, maintaining the morale and physical well-being of soldiers, their families and dependents.

Soldiering will be remembered for paragraph 103 which codified the Military Covenant. However, the principle of reciprocity recurs throughout the paper, with references made to the obligations between nation, Army and individual soldier. Roberts summarises:

> I wanted to get across the point that if the soldier has unlimited liability, so must the nation have it … I just want – and I think we got there – it to be widely understood, that soldiers potentially give everything. In return, the nation, private or public, must be prepared to give everything back.

The Moral Component Updated

Military doctrine is regularly updated, reflecting not just the experience gleaned from recent operations, changes in the global strategic environment and public policy, but also the Army's institutional mindset. The latest doctrine supersedes earlier publications. For example, ADP *Land Operations* (2005) made redundant the previous five-volume edition of capstone doctrine: it included a chapter on the Moral Component, with a paragraph entitled 'The Military Covenant'.[114] Although *Soldiering* attempted to convey the verities of being a soldier, there is a sense that concept was institutionally demoted. Conversely, *Values and Standards*, written for every soldier was re-issued in 2008.[115]

While the Army's core values and standards are unaltered, the 2008 document is radically different in tone from its predecessor, underlining the evolution of all military doctrine. It takes a far less judgmental approach to civilian society, while any claim to a 'right' to difference is dropped:

> It is operational effectiveness that requires the Army to have values and standards that are different from society – 'need to be different' not the right to be 'different' [*sic*].[116]

113 Ibid., paragraph 203.
114 ADP, 2005, paragraph 715.
115 ADP, 2008.
116 ADP, 2008, paragraph 2.

Equally, there is no suggestion that the Army is, or should be, a special case in terms of human rights and diversity legislation. However, it emphasises that the Army is different from civilian employers: 'Our Values and Standards apply at all times: whether on operations, in barracks, in our homes or off duty'.[117] In addition, by 2008, the civilian and military view of the Military Covenant had diverged. Civilians saw the Covenant, predominantly, as a bilateral compact between Armed Forces personnel and the government; according to *Values and Standards*, the Covenant was between Army and soldier:

> The Army's loyalty to the individual [soldier] is expressed in the Military Covenant – it manifests itself in justice, fair rewards and life-long support to all soldiers.[118]

Although *Values and Standards* has a 'moral requirement and a functional utility',[119] this aspect of the Moral Component was revisited, in part, because of the abuse of Iraqi civilians by British soldiers, underlining that doctrinal theory does not necessarily accord with operational reality.[120]

In 2005, *Soldiering: The Military Covenant* was subsumed into the capstone *Land Operations* and revised. Paragraph 103 was amended, becoming far more equivocal about what soldiers may expect in return for their service and sacrifice:

> So, *at the very least*, British soldiers *should always be able to expect* the Nation, and their commanders, to treat them fairly, to value and respect them as individuals, and to sustain and reward them and their families with *appropriate* terms and conditions of service.[121] (my italics)

The use of the conditional – 'should always be able to expect' – rather than the 2000 imperative – 'must' – implies a dilution of expectation. Similarly, 'the common bond of identity loyalty and responsibility' has, in 2005, only sustained the Army throughout its history 'to a greater or lesser extent'. In a departure from the original, the revised paragraph spells out what happens if one of the Covenant's three agents should fail to honour its terms:

> Unless Nation, Army and soldier alike recognize and understand that it must be upheld come what may, then it fails. If it fails then first goodwill and then, ultimately, trust, is withdrawn.[122]

117 Hackett, The Lees Knowles lecture, cited by ADP, 2008a, paragraphs 7, 3, 34.
118 Ibid., paragraph 13.
119 Ibid., paragraph 7.
120 Chapter 7.
121 ADP, 2005, paragraph 715.
122 Ibid.

Although the Covenant may be a morally-based understanding, the cost of failure is heavy. Mutual trust, as General Dannatt reminded all soldiers in 2008, is integral to the Army's worldwide reputation for excellence.[123]

The post-migration, civilianised Covenants that began to emerge in 2007 were far from the spirit of the Military Covenant codified in Army doctrine. Former Brigadier Philip Mostyn, one of those involved in drafting the revised Moral Component of 2005, says that the original, underscored by its use of the term 'covenant', was 'designed to appeal to emotion not reason'.[124] He adds:

> In requiring the nation to 'treat them [soldiers] fairly, to value and respect them as individuals and to sustain and reward them', it requires no more than any good employers provides in return for the ultimate sacrifice. So as a transaction it does not withstand close rational scrutiny.[125]

He observes that both versions share ambiguities, not least in the phrase, 'It is a Covenant, not a contract and it is binding in every circumstance'. In addition, Mostyn queries the status of the Army as a component agent:

> In practice, 'the Army' means some Generals, who are themselves also soldiers. So they are on both sides of the 'deal' at the same time. All of this suggests that the Covenant is not so much a deal as simply the Army's hope and plea that the loyalty which soldiers give to those who send them to war is returned in kind.[126]

A hope and plea is very different from the post-migration demands made by various civilians groups to give statutory recognition to their individual interpretations of the Covenant. In emphasising the transactional, the concept's civilian advocates unwittingly undermined the Covenant and challenged the Army's ethos of service.

Conclusion

Military doctrine constantly evolves. While changes to tactical and operational doctrine based on technological advance are easy to quantify, those relating to ethical doctrine are more nuanced. As *Soldiering* states: 'Other aspects of Fighting Power such as morale, doctrine and leadership do not easily lend themselves to objective measurement'.[127] However, ethical doctrine should not be subject to too much change. 'The brightest of all realities which we call the good' was perceived

123 ADP, 2008a, Foreword.
124 Mostyn, telephone discussion.
125 Ibid.
126 Ibid.
127 ADP, 2000a, paragraph 4–1.

by Plato to be immutable and eternal.[128] The six core values identified by the Moral Component would be universally accepted as positive moral qualities, which in the doctrinal context are imbued with functional as well as moral significance.

The maintenance of morale is a Principle of War, second only to the selection and maintenance of the aim. Ensuring that soldiers 'are sustained and rewarded by commensurate terms and conditions of service' is surely a key factor in maintaining that morale. With the codification of its ethics in general and the Military Covenant in particular, the Army set down exactly what it expects from its soldiers: equally, soldiers now know exactly what they can expect in return from the Army and the nation. In some respects, the ethical doctrine resembles the Commandments: it lays down absolutes. While this approach could be applauded for its attempts to build a moral community,[129] it could also be criticised for its lack of realism and flexibility.

The Army's codification of ethical doctrine was entangled with concurrent changes to human rights and equal opportunities legislation imposed upon it from outside. The apparent siege mentally of the 1990s has given way to acceptance and a more flexible approach to civilian society. *Soldiering* has been described as a 'creed' by its writer and 'an essay' by another senior officer. Many interviewees have described the Military Covenant being an imperfect codification of a previously tacit, undefined concept. With the Moral Component the Army set itself up an exemplar, created an ethical paper trail and formalised the moral obligations between Army, soldier and nation – along with the expectation that those obligations would be discharged.

---

128 Plato (trans. Lee), 1987, p. 322.

129 Chapter 3.

Chapter 3
The Contractual Context

Introduction

In March 2006, the parents of the four young soldiers whose deaths at Deepcut had prompted the Blake Inquiry declared: 'Our children signed to serve the country. It is time the country served them'.[1] The year-long Inquiry was one of more than a dozen investigations into the circumstances surrounding the deaths that occurred between 1995 and 2002 at the Princess Royal Barracks, which led to appraisals both inside and outside the Army about the treatment of recruits and trainees, the Army's attitude towards bullying, as well as its obligations – both legal and moral – regarding its duty of care; i.e. the responsibility to protect health, safety and well-being. Emphasising the Army's obligations to soldiers, the Blake Report was among the first references to the Military Covenant by a civilian:

> The discharge of the mutual commitments in the Military Covenant should not start with a 'credit note' to explain inadequate staffing ratios, recreational and welfare facilities.[2]

In 2008 in the High Court, Blake, by then Mr Justice Blake, would rule on the settlement rights of Gurkha veterans who had served the Crown before 1997.[3] The Covenant was central to his judgment.[4] In 2006, Blake emphasised the Army's obligations to soldiers and trainees in the context of the Covenant; in 2008 he underscored the nation's duty to its soldiers – particularly those who had been in combat.

The Military Covenant highlights the distinction between moral and legal contracts, which is the key theme of this chapter. Aristotle underlined the delineation between moral and legal spheres: 'Now it seems that, as justice is of two kinds, one unwritten and the other legal, one kind of friendship of utility is moral and the other legal'.[5] A covenant is an archaic and comparatively open-ended concept, dating back to the Bronze Age: consequently, its relevance to a twenty-first-century organisation such as the Army is, ostensibly, far from obvious.

---

1 BBC, Deepcut Families Vow to Fight On, 29/06/06.

2 Blake, 2006, p. 391.

3 *R (on application of Limbu and others) v Secretary of State for Home Department*, EWHC 2261, 2008.

4 Chapter 7.

5 Aristotle (trans. Ross, revised Ackrill and Urmson), 1980, p. 216.

However, in exploring the covenant-contract dichotomy, the utility of a covenant emerges. This chapter aims to establish why the Army developed the concept of the Military Covenant, and, in doing so, what it intended to communicate both to soldiers and civilian society.

Although defined as 'a written or spoken agreement between two or more parties, intended to be enforceable by law',[6] contracts exist in many different forms, but share common norms. Intertwined with them are concepts of trust and promises – those obligations that, like the Military Covenant, have moral if not legal force. Legal contracts usually afford protection, as well as a means of recompense if broken. Similarly, the legal concept of 'limited liability' provides protection against unlimited loss. However, despite all the constraints that are imposed upon them, soldiers do not have a legal contract of employment. This anomaly is all the more curious in the light of Hackett's observation that all soldiers have committed themselves to the 'contract of unlimited liability'.[7] The Covenant's absence of tangible sanction and recompense should its terms not be fulfilled still holds despite the concept being given statutory recognition in 2011. This chapter examines nature of trust and the risks inherent in trust. In 2005, Army doctrine stated that were the Covenant not upheld, 'first goodwill and then, ultimately, trust is withdrawn'.[8] CGS Dannatt stated that trust is the foundation of British soldiers' global reputation for excellence and the 'lifeblood that sustains the Army'.[9] Violations of trust within any workplace can destroy the so-called psychological contract, an entity that is also examined.

The Contractual Context

A covenant is a promissory construct that is rooted in the Judeo-Christian religious tradition. The Divinely granted Covenants – with Noah, Abraham, Moses, David and at the Last Supper – bestow by association a quasi-mystical authority on the Military Covenant, elevating it above a mere workplace agreement.[10] Elazar observes: 'In its heart of hearts, a covenant is an agreement in which a higher moral force, traditionally God, is either a party to or guarantor of, the particular relationship'.[11] Originally a 'theo-political' concept,[12] it is a specific form of agreement, which differs from a charter, compact or contract, the latter of which today has, primarily, legal associations. Many Biblical covenants were more political than theological in character, establishing lines of authority, distributions

6 *Shorter Oxford English Dictionary*.

7 Hackett, 1983, p. 208.

8 ADP, 2005, paragraph 713.

9 Dannatt, ADP, 2008, Foreword.

10 Exodus 19 1–8: Exodus 24: Matthew 26.

11 Elazar, 1996, p. 2.

12 Ibid., p. 2.

of power, bodies politic and systems of law.[13] Some were pacts between individuals such as the covenant between David and Jonathan. However, interpretations of 'covenant' vary, with different meanings 'in law, politics and the Bible', according to theologian James Torrance, who notes that Scottish law refers to a 'deed of covenant' and a 'marriage contract'.[14] In 2005, military doctrine stated that the Military Covenant was '... a Covenant and not a contract and it is binding in every circumstance'.[15] It can be inferred that the Army wished to convey that, although the Covenant is not a *legal* contract, it must be upheld.

The utility of a covenant to the Army emerged when the Moral Component was under consideration in the latter half of the 1990s. Archaic, unfamiliar to many and associated with religious faith, a covenant is a curious concept for a contemporary secular institution to adopt. Highlighting the utility of covenantal agreements to the secular sphere, (Rabbi Lord) Sacks suggests:

> Covenant occurs when two individuals or groups, different perhaps in power, but each acknowledging the integrity and sovereignty of the other pledge themselves in mutual loyalty to achieve together what neither can achieve alone.[16]

The Mayflower Compact reflects the mid-sixteenth-century revival of the covenantal principle as the basis for organising society among Protestant groups. Signed by pilgrims at Cape Cod in 1620, it became a governing document for the Plymouth colony. It begins:

> In the Name Of God Amen. We, whose names are underwritten ... do by these presents, solemnly and mutually, in the Presence of God and one another, covenant and combine ourselves together in a civil Body Politick, for our better ordering and preservation. ...[17]

Torrance suggests religious dissent and political radicalism often coalesced, with the term covenant becoming a 'revolutionary symbol' in, for example, Civil War-era Scotland.[18] The anti-Home Rule Ulster Covenant of 1912 – a legacy of the Presbyterian roots of many in the province and an echo of the 1638 and 1643 Scots' Covenants – was signed by almost a quarter of a million men 'humbly relying on the God whom our fathers in days of stress and trial confidently trusted'.[19] South Africa's Day of the Covenant commemorated the vow to God to build a church in exchange for victory that the Vortrekkers made before the Battle

13 Elazar, 1996, p. 2.
14 Torrance, 1970, pp. 51–76.
15 ADP, 2005, paragraph 713.
16 Sacks, 2002, p. 151.
17 H.W. Jones, 1975, pp. 43–57.
18 Torrance, 1981, pp. 225–243.
19 The Ulster Covenant.

of Blood River in 1838.[20] Akenson argues that Afrikaaners, Ulster Presbyterians and Israelis share a similar cultural perspective; with their belief that they belong to covenanted nations, they are among the Elect or God's chosen people.[21] Today, Israel's territorial claims are justified in part by the Abrahamic Covenant later confirmed at Sinai, which continues to lend them uncompromising weight. Although an ancient concept, covenants have had continuous relevance for some faith-based communities for centuries.[22] As Akenson observes, 'the Hebrew covenantal structure has lasted from the middle of the Bronze age to the present'.[23]

Covenantal agreements, including the Military Covenant, can be entered into which are entirely secular, yet they remain different from other types of contract. In 1919, the Covenant of the League of Nations bound members – 'high contracting parties' – to 'the acceptance of obligations not to resort to war'.[24] In 1946, the League and its Covenant were superseded by the United Nations and its Charter. Although both documents set out the identical nature of the organisations' fundamental objective – the maintenance of international peace and security – a charter and a covenant are two very different instruments: 'the word charter customarily denotes the granting or gaining of rights, powers or functions'.[25] The Charter introduced new principles into international law including, according to Moore and Pubantz, 'the nonresort [*sic*] to war', which was set out in Article 2. In contrast, the League's Covenant 'essentially suggested a voluntary commitment not to resort to force'.[26] While the provisions of the Charter were more comprehensive than the Covenant, for example enshrining the principles of self-determination and introducing mandatory forces for peacekeeping, the Charter's chief importance is that, in the post-Cold War context at least, it has bolstered the authority of international law, particularly in relation to any form of armed intervention. While striving towards the same objective, the foundation documents of the two organisations typify how a covenant, even one drawn up for political purposes, is different from a charter.

The Military Covenant emphasises the distinction between a trust-based covenant and a contract, which usually has a legal basis. Covenants are 'outside the world of contracts and exchange' and the parties to them are bound by moral reciprocity.[27] Elazar suggests: 'A covenant differs from contract in that its morally binding dimension takes precedence over its legal dimension'. James Torrance observes that in the religious context, 'the God of the Bible is a Covenant-God and not a contract-God'. Consequently, Grace is not conditions-based:

20 16 December is now Reconciliation Day.

21 Akenson, 1992.

22 Excluding 'legal' covenants concerning property.

23 Akenson, 1992, p. 356.

24 *Covenant of the League of Nations.*

25 Moore, and Pubantz, 2002, p. 57.

26 Ibid., p. 56.

27 Sacks, 2002, p. 77.

> A covenant brings its promises, its obligations and its warnings … In the Bible, the form of the covenant is such that the Indicatives of Grace are always prior to the obligations of law and human obedience. 'I am the God of Abraham, Isaac and Jacob, I have loved you and redeemed you and brought you out of the land of Egypt, out of the house of bondage, therefore keep my commandments'. But legalism puts it the other way round.[28]

To paraphrase Torrance's point, God makes a covenant at Sinai with the Israelites unconditionally; there is no negotiable 'If you do x, I'll do y'. He does, and will continue to do x, whatever happens. For Iain Torrance, a covenant is 'a binding agreement reached without conditions:'[29] x is not dependent upon condition y, upon which most legal contracts are based. Naturally, all parties of a covenant are beholden to uphold it, but the pressure to do so is exerted not by legal obligation but by moral force, primarily that of keeping a promise. Unlike a breach or non-fulfilment of a legal contract, if one of the agents breaks the terms of a covenant, there is no opportunity for any restitution, other than moral sanction.

The covenant concept was brought to collective military attention indirectly by Rabbi Sacks, and more directly by Army chaplain, Iain Torrance, through the offices of Chaplain-General Victor Dobbin. He arranged for Sacks to give a lecture 'Leadership, Vision and Values' to senior commanders at Sandhurst and also commended the Rabbi's writings to them, among which were explorations of covenanted communities.[30] This coincided with an emerging sense that, following a series of scandals and adverse publicity concerning bullying and allegations of racism, the Army should be giving greater consideration to the Moral Component. Adjutant-General Sir Michael Rose arranged a symposium at Bagshot Park in 1995 at which it was reviewed. This included a presentation on the attitudes of recruits, along with a lecture by Iain Torrance, who subsequently wrote a paper on ethics and the military community, in which covenants are explored.[31]

Through the Moral Component, the Army asserted its difference from civilian society, a development encouraged by members of the Army's Chaplaincy. Informed by the work of Alastair MacIntyre, Lesslie Newbigin and Michael Polanyi that address the fragmentation of moral authority in Western culture, Dobbin advised senior commanders to consider what sort of community the Army should be.

> Something was wrong and needed to be put right. Could the Army be a moral community but not a cult? While some might want the Armed Forces to reflect

28 Ibid.

29 Iain Torrance, 1998, p. 4 (nephew of James).

30 Dobbin, discussion.

31 Iain Torrance, 1998.

> society, the values and standards of society might not be appropriate. Could the Army choose its own values and standards?[32]

With its history and traditions giving it a sense of cohesion, Dobbin considered the Army ideally placed to develop into the sort of rational, values-led community advocated by MacIntyre et al. He anticipated that soldiers would support and assimilate the Army's articulation of its core values because of their pride in being a soldier and in their regiment. In addition, by developing, articulating and inculcating its values, not least because of the imperative of operational effectiveness, the Army reinforced its difference from civilian society, as well as from the civilian organisations within it which also employed 100,000 people.[33] Above all, the assertion of the Army's right to difference was made in the context of the deliberate pursuit of finding the answer to the 'moral complexities within its own community'.[34] This right to difference was not the right to be granted exemption from legislation concerning minorities and human rights, nor the right to condone bullying and racism, all issues for which, at the time, the Army was being castigated by the civilian sphere. Although some may have considered any talk of 'rights' an over-assertive approach, and one that could be interpreted as an expression of institutional moral superiority, Dobbin maintains that the institutional intention was benign. The Army's right to be different was essentially its right to develop a corporate moral framework, which came to be represented by its self-selected values and standards. Because of demands military operations, this process involved rejecting some external, civilian values, such as the primacy of the individual. Dobbin says: 'I was definitely advising the Military of its right to be different'.

Soldiering: The Military Covenant stemmed from the Army's inquiry into its right to difference. In the 1990s culture clash between the military and civilian spheres, the Army came off worse. However, many senior commanders, including those involved in developing the Moral Component consider that, in the context of the promotion of military effectiveness and soldiers' service, the assertion of institutional difference – in relation to values, the law or practice – was legitimate. For Sir Michael Rose, the Army's difference from civilian society was based upon its ethos: 'a willingness by individuals to subordinate their interests to the common good – even to the extent of sacrificing their lives'.[35] With the development of the Moral Component, the Army's 1996 *Right to be Different* paper was superseded. Rose says:

> In its place came the idea of a Military Covenant in which the Government would recognize the different demands on soldiers and would accept a special

32 Dobbin, discussion.

33 Ibid.

34 Ibid.

35 Rose, email to author.

> responsibility to look after them that went beyond its duty of care for the civilian population.[36]

With the Moral Component, the Army retreated from its formal claims to a 'right' to difference – and consequently may be perceived as being vanquished by the progressive forces of civilian society. However, the Military Covenant encapsulated and promoted the 'unique nature' of military service, predicated on the Army's ethos and soldiers' sacrifice. The Covenant conveyed the difference of the military sphere in a manner that was acceptable to civilian society, while underlining society's moral responsibilities to the military.

The utility to the twenty-first-century British Army of a covenant, an agreement which can be found in the Old Testament, is far from obvious. For Iain Torrance, the paramount factor that sets a covenant apart from other contracts is that, integral to it, are a transcendent reference and commitment. Torrance argues that for soldiers, their covenant is 'to serve'.[37] Once this central commitment of a covenant is in place, the other factors that can characterise the concept – perpetuity, equality, voluntary consent, unconditionality – fit around to make the whole. The covenantal concept conveys the unique nature of military service with its inherent sacrifice – and the corresponding obligation owed to soldiers. For Torrance, the individual soldier's preparedness to make the ultimate sacrifice is its lodestar:

> I did indeed see that *offer of sacrifice* by the soldier as the root of 'covenant', to which a grateful public responds. It is a response of gratitude, not one from within a contract.[38] (italics, *sic*)

The archaic, Biblical associations of the covenantal concept implicitly reinforce the Military Covenant's moral status. However, its moral force derives from soldiers' service and potential sacrifice.

The Moral Component was developed for the Army, by the Army, but with a mindfulness of an external audience in civilian society. Consequently, while the Military Covenant had military utility – strengthening the bonds between the chain of command, and between the Army and the individual soldier – as a contractual agreement it also had practical utility. In short, it underlined soldiers' unlimited liability without making the Army or the MoD legally liable for it. Although the Covenant codified the duty of care to soldiers – and parenthetically their families – in terms of 'fair treatment' and 'commensurate terms and conditions of service', this was a moral rather than a legal duty. *Soldiering*'s author says:

> That whole discussion about the terminology 'contract' and 'covenant' occurred – because those of us who were talking about it, were very conscious

36 Ibid.

37 Iain Torrance, 1998, p. 27.

38 Email to author (his italics).

> that we needed to say something to the soldier – and the soldier's family – about what to expect in return. We knew that if we used anything that linked to the contract, it simply wouldn't pass go. When the Army Board looked at it, the Second Permanent Under-Secretary would stand up and say, 'we can't possibly agree to a document which says, 'the soldier writes a blank cheque'; you're saying if it's contractual, the nation must write a blank cheque in return.[39]

Major-General (retired) Andrew Ritchie agrees, concurring that 'contract' would never have been approved by the MoD. He likens the term 'covenant' to 'dominions' that described some self-governing nations within the British Empire: almost obsolescent, opaque but useful.

> From what I remember it started as the Military Contract, but I personally felt very strongly contract was the wrong term. We don't have contracts in the Army. It's not a term we ever use. We don't have an employment contract, or at least not one that anyone would recognize, so let's be careful about saying there's a contract.[40]

By underlining the delineation between the moral and legal duty of care, the Covenant codifies the moral obligation to soldiers while avoiding any legal liability.

The Military Covenant reconciles soldiers' service and sacrifice with the possibility that it might not be rewarded. Both the 2000 codification and its 2005 successor state that in exchange for 'personal sacrifices', soldiers 'must (or, in 2005 'should') be able to expect' fair treatment, respect, to be valued and 'to be sustained and rewarded by commensurate ('appropriate') terms and conditions of service'. The term 'should be able to expect' is very different from 'will receive', which would connote a conditions-based arrangement. (If individual soldiers do x, the Army and the Nation will do y for them in return.) The 2005 Military Covenant states: 'Soldiers are bound by service'.[41] That service involves sacrifice, but 'the nature of service is inherently unequal: soldiers may have to put in more than they receive'.[42] This codified warning to soldiers came two years after the start of combat operations in Iraq, when the implications of their contract of unlimited liability must have been apparent. However, such is the nature of the soldiering, soldiers will continue to serve the nation, despite the nation or the Army not fulfilling their moral obligations to them. While this could cause bitterness, especially among veterans, as a contractual concept, the Covenant captured and reframed this essential fact – probably unwelcome, but not historically inaccurate – of what it is to be a soldier.

39 Roberts, interview.

40 Ritchie, interview.

41 ADP, 2005, paragraph 0713.

42 Ibid.

The Military Covenant alerts the nation to its responsibilities and obligations to soldiers, justified by the moral imperatives of the unique nature of military service. Although it is not legally binding upon the nation, the nation cannot shirk those responsibilities. According to Roberts:

> If I intended anything, it was to have something written down, which told us what not only the responsibilities of us soldiers were, but the responsibilities of those who sent them. That was the most important thing. It was to tell the truths about soldiering, to everybody who was responsible for what the soldiers did. The individual soldiers, their commanders, but also society, the politicians, the civil servants, the media.

While no Permanent Under-Secretary would endorse writing blank cheques to soldiers, Roberts wanted to convey that soldiers did exactly that for the nation. He sees the relationship between the individual soldier and the nation mediated by the Army, but says it behoves all parties to keep it in balance. However, CGS Dannatt stated the relationship between the military and the Government was 'out of balance'.[43] Despite this, soldiers carried on serving. If the relationship between them, the Army and the nation had been formalised on a basis that was legally rather than morally contractual, as is the case in other professions, they might have had justifiable grounds for claiming compensation or withdrawing their labour.

One criticism of the Military Covenant is that it is a contract involving the nation, about which the nation was not consulted.[44] An officer involved in writing the 2005 Moral Component, Brigadier (retired) Philip Mostyn observes: 'We wrote a contract that the Government never signed up to and used it as a stick to beat them with'.[45] The 'broken' Military Covenant, a phrase used by the Labour government's critics if not by senior commanders, did indeed become a partisan weapon from late 2007. As much as the Army shies away from legal contracts, those involved in practical policy-making, such as ministers and civil servants, might want to avoid the messy entanglements of moral contracts. Indeed, the Ministry of Defence was criticised by Blake for focusing too narrowly on the legal aspect of its duty of care, at the expense of its moral responsibilities. However, the dogged refusal to acknowledge a moral concept that was accepted and supported by the public left Ministers looking flat-footed. The Social Contract itself, upon which much of Western polity is based, was never assumed to have gained formal consent from those emerging from Hobbes' state of nature. It is a hypothetical construct: individuals cede their natural rights and liberties to a sovereign authority in return for a peaceful, just society. Consent was assumed: no one was asked to sign on the dotted line. Similarly, before migration, the public's consent to the

43 Dannatt, interview.

44 Raised by Dr John Stone.

45 Mostyn, discussion.

Military Covenant was assumed: after migration it was actualised, reflected by the support for military charities.

Most soldiers only became aware of the Military Covenant once it had migrated to, and was championed by, the civilian sphere. Between 2000 and late 2006, the concept was institutionally unimportant, with the Army prioritising its self-selected values and standards. Half a century ago, General Sir John Hackett had his own vision of why values and standards were essential:

> Thus while you may indeed hope to meet these virtues in every walk of life and a good deal of educational effort is spent on developing them as being generally desirable, in the profession of arms they are functionally indispensable. The training, group organizations, the whole pattern of life of the professional man at arms is designed in a deliberate effort to foster them, not just because they are morally desirable in themselves, but because they contribute to military efficiency.[46]

For Hackett, civilian and military virtues were identical, but in the military they were an operational imperative; 'jewels set in blood and iron'. While it appeared to be taken for granted that soldiers needed virtues to be professionally effective, some interviewees are wary that the process of codification might triumph putting those values into practice. Ritchie suggests: 'It could be compared with our unwritten Constitution. We don't really do this stuff that's written down'.[47] Mostyn is unconvinced about the usefulness of codification, pointing out that Hitler's SS had a code of honour:

> Of course it provides a public framework for laying down boundaries but does it change behaviour? Are we sure that soldiers weren't carrying the card[48] while robbing the shop? Human behaviour is not greatly affected by these sort of codifications. They do not make much difference.

How much difference the codification of the Moral Component has made to the ethics of the individual soldier is difficult to quantify. CGS Dannatt saw the abuse of Iraqis at Camp Breadbasket as a failure to inculcate the Army's values and standards, echoed by the 2008 Aitken Report.[49] However, Major-General (Retired) Stephen Andrews, suggests that the Moral Component promoted a 'growing awareness of values-led leadership' in the Army.[50]

While professions such as medicine are characterised by their long-standing codes of ethical practice, the development of the Moral Component can be seen as part of a wider 1990s' national trend towards the codification of rights and

46 Hackett, 1962.
47 Ritchie, interview.
48 All soldiers were given a credit card-sized card listing the six core values.
49 Aitken Report, 2008, Foreword.
50 Andrews, interview.

responsibilities in both the public and private sectors. An example is the Charter initiative, beginning with the Citizen's Charter, which was to apply to areas of the public sector, including local authorities. In 1994, the Nolan Commission set out the 'Seven Principles of Public Life', to which public office-holders should adhere. It recommended that 'all public bodies should draw up Codes of Conduct incorporating these principles', which included integrity and selflessness.[51] In addition, many organisations began to develop their own ethical component with the emergence of Corporate Social Responsibility (CSR), which often focused upon environmental ethics. Mostyn suggests that the development of the Army's Moral Component was in part an attempt to keep up with the trend. 'A lot of industry was saying we must have "articulated values"; everybody was doing it'. Many view CSR – especially in connection with environmentalism – with a degree of scepticism, considering it as nothing more than a coat of 'greenwash' by some corporations keen to enhance their public image.[52] In the commercial sector, an ethical code is not seen as intrinsic to organisational effectiveness in the same way that it is perceived to be by the Army. With the delegation of military responsibility through Mission Command and the rise of the so-called 'strategic corporal', soldiers bear a greater individual responsibility to act in a morally and ethically acceptable manner. However, this is less to do with protecting the Army 'brand' than functional necessity; that is, promoting operational effectiveness.

Covenants are contracts founded upon trust rather than legal obligation. The trend for codifying ethics and developing CSR within civilian society has, paradoxically, coincided with an erosion of trust in many professions and institutions. For O'Neill, the advent of the 'accountability culture' in Britain's public sector ever since the late 1980s, has led to the reputation and performance of public sector workers and professionals being doubted, undermining both professional judgment and institutional autonomy.[53] Historically, professions overrode the dictates of the market place: as Burk suggests 'relations with clients were based on the principle of trust (not caveat emptor), and professionals were expected to perform their duties even if that required an element of self-sacrifice'.[54] The Army is not the only profession whose 'professional space' has been encroached upon by successive governments: senior commanders should not be castigated for this any more than doctors, academics or MPs.[55] However, while worthy in principle, transparency and codification does not guarantee that trust is engendered: as O'Neill points out, some regimes of accountability may damage rather than reinforce trustworthiness: 'Trust seemingly has receded as transparency

51 Nolan, Summary, 1994.

52 Greenpeace USA urges corporations to 'Clean up your act NOT your image'.

53 O'Neill, 2002.

54 James Burk, Expertise, Jurisdiction and Legitimacy of the Military Profession, in (eds) Snider and Watkins and (ed.) Matthews, 2002, p. 24.

55 Forster argues that senior commanders have not defended their 'professional space'.

has advanced'.[56] According to Sacks, the market and the state have weakened trust-creating institutions. Like others, O'Neill examines the contemporary 'crisis of trust', which, according to Seldon, appeared to become acute in Britain in 2009 following the banking collapse and the Parliamentary expenses scandal.[57]

Integral to social relations, trust is pervasive but only considered when it is threatened or destroyed. A 'functional prerequisite for the possibility of society' and 'indispensable to social relationships',[58] trust is integral to the subject-matter of the humanities, economics, law and social science disciplines, but seems curiously overlooked by them. The concept retains a certain miasmic quality. In his examination of the philosophy of trust, Bailey argues trust is elusive to philosophers, with many simply ignoring or presupposing it,[59] just like the rest of us. Baier notes that neither trust nor trustworthiness are among the virtues Plato dwells on as necessary in his Republic.[60] To trust, above all, is to take a risk: 'the vesting of confidence in persons or abstract systems, made on the basis of a "leap into faith" which brackets ignorance or lack of information';[61] trust is 'where there is some, perhaps substantial risk, of default or untrustworthy behaviour';[62] 'trust is a bet about the future contingent action of others'.[63] Trusting makes us vulnerable: 'all trust risks disappointment';[64] 'placing trust means acting as if the risk were non-existent'.[65] Trust is crucial at a macro-social level, as well as to interpersonal relations. Credit, from Latin *credere*, relies upon it. Giddens argues that the abstract systems of late modernity – such as the globalised monetary system – depend on trust to create large areas of relative security for the continuance of day-to-day life.[66] As Baier suggests we inhabit a climate of trust as we inhabit an atmosphere, noticing it as we notice air, 'only when it becomes scarce or polluted'.[67]

Trust differentiates covenants from legal contracts. While legal contracts are intrinsic to civilian society, they can be correlated with a decline in trust and evidence of social dysfunction.[68] As Beale et al. suggest, contract law developed in parallel with nineteenth-century enterprise and is primarily concerned with supporting exchange.[69] Although the law does not govern most if not all of the small, often tacit, exchange agreements that are part of day-to-day life, the withering of

56 O'Neill, 2002, p. 43.
57 Seldon, 2009, p. x.
58 Lewis and Weighart, 1985, pp. 967–986.
59 Bailey, Open University, BBC.
60 Baier, 1986, pp. 231–260.
61 Giddens, 1991, p. 244.
62 Hardin, 2006, p. 101.
63 Sztompka, 1999, p. 25.
64 O'Neill, 2002, p. 24.
65 Sztompka, 1999, p. 31.
66 Giddens, 1991, p. 133.
67 Baier, 1986, pp. 231–260.
68 Fukuyama, 1995; Seldon, 2009.
69 Beale et al., 2008.

trust in both Britain and the United States that has occurred in recent decades towards many institutions and professions, 'imposes a kind of tax'.[70] Fukuyama suggests legal apparatus is a 'substitute for trust'.[71] However, while Hardin argues a 'massive system of contract enforcement' enables agents to enter into very complex cooperative relations that are mutually beneficial,[72] Kay suggests that in the commercial sector, legally enforceable transactional agreements only add to profitability in the short-term; in the longer term, they erode the trust and goodwill that are a vital part of the relational 'unwritten' contracts between those who are party to legal contracts:

> Lawyers for American companies spent hundreds of billable hours drawing up contracts to which no one ever referred. Their Japanese counterparts engaged in complex business relationships with no formal agreements at all, or ones that covered just a single sheet of paper.[73]

This can be attributed to Japan's 'high trust' culture, identified by Fukuyama. According to Macaulay, who with Macneil developed the relational theory of legal contract, the 1960s Empire State Building sale involved more than 100 attorneys and resulted in a contract of more than 400 pages, which tried to cover every eventuality.[74] Fukuyama argues that while contract and self-interest are important, the most effective organisations are based on communities of shared ethical values. Such communities do not require extensive contracts and legal regulation because prior moral consensus gives members of the group a basis for mutual trust.[75] The necessity of the high trust relationship between soldiers suggests that legal contracts between them would be of little relevance to combat effectiveness.

The development of a comprehensive ethical code such as the Moral Component whose most important aspect, ten years on, is the trust-based Military Covenant sends out a message that the Army is indeed different from civilian society. Covenants are 'outside the world of contracts and exchange' and the parties to them are bound by moral reciprocity.[76] While this sort of exchange arrangement might have suited tribal societies in the Bronze age, they may not be the most practical arrangement today. Rousseau argued that laws of natural justice – dependent upon reciprocity and lacking sanction if broken – were ineffectual; 'such laws merely benefit the wicked and injure the just, since the just respect them while others do not do so in return'.[77] While acknowledging the covenantal

70 Fukuyama, 1995, p. 27.

71 Ibid., p. 29.

72 Hardin, 2006, p. 59.

73 Kay, *Financial Times*, 16 March 2010.

74 Macaulay, 1963, pp. 55–67.

75 Fukuyama, 1995, p. 26.

76 Sacks, 2002, p. 77.

77 Rousseau (trans. Cranston, 1968), p. 80.

concept and its principles of voluntarism, trust and reciprocity, Social Contract theorists imply that, because of the absence of sanction, such agreements were inadequate compared with a body of civil and criminal law that must be an integral part of a modern state. Unlike soldiers, few other professionals would rely upon a covenant to ensure they were properly cared for in case of a workplace accident. Although the decline of trust is lamented by many, including Fukuyama, who deem the parallel rise in litigation a social negative, perhaps senior commanders in the late 1990s were too trusting that the nation – both policy-makers and the public – would keep its side of the Covenant bargain.

Although the Moral Component was written before the advent of 'Blair's Wars', it is paradoxical the Army should have turned its back on legally-based contracts setting out more decisively soldiers' terms and conditions, advocated, in part, in 1995 by Bett.[78] This legal lacuna is all the more curious in the context of the ever-increasing legal constraints on British soldiers, whether military law, civilian legislation, international humanitarian law, the Geneva Conventions, the Law of Armed Combat, or the tenets of *jus in bello*. However, Roberts suggests, the body of formal regulation was not wholly adequate to convey the magnitude of soldiers' responsibilities. As *Soldiering* states:

> All British soldiers must discharge their duties not just according to orders and law, but consciously and clearly for the greater good.[79]

The Moral Component was, in part, a riposte to, and defence against, the development in the late 1990s of what this study identifies as the legal component; that is, civilian legislation that could have an adverse impact on the Army, particularly concerning employment practice, human rights and health and safety. As such, it was a subtle but effective weapon: few other institutions could get away with couching special pleading in the guise of a quasi-sacred agreement such as the Covenant, with all its atavistic appeal.

The Military Covenant encapsulates the distinction between the legal and moral agreements. The concept not only makes explicit the mutual obligations of the three agents – soldier, Army and Nation – but stresses that, similar to a promise, it is 'binding in every circumstance'. Consequently, the Covenant imposes a moral obligation on the individual soldier, the chain of command and the nation to honour its terms. While there are no legal penalties for not doing so, according to the 2005 formalisation there are moral penalties, specifically the withdrawal of goodwill and trust.

---

78 Michael Bett, 1995. Independent Review of the Armed Forces' Manpower, Career and Renumeration Structures. London: HSMO.

79 ADP, 2000a. Army Doctrine Publications. Ministry of Defence. © Crown copyright 2014.

> Unless Nation, Army and soldier alike recognize and understand that it must be upheld come what may, then it fails. If it fails then first goodwill and then, ultimately, trust, is withdrawn.[80]

Trust is vital to the British Army and central to the Moral Component. Doctrine states that any breakdown in trust has a 'corrosive effect on operational effectiveness'.[81] CGS Dannatt stated that the British Army's worldwide reputation for excellence 'derives from, and depends upon, unequivocal commitment, self-sacrifice and mutual trust'.[82] Similarly, 'social misbehaviour can undermine trust and cohesion and, therefore, damage operational effectiveness'.[83] The United States Army relies on trust 'the way the human body relies on blood'.[84] Clearly, the trust that is the essential foundation for cohesion and commitment within a combat unit on the battlefield is of a different order of magnitude than that necessary between colleagues in a more typical workplace. Major-General (Retired) Stephen Andrews says:

> The most important part of the Military Covenant is the trust soldiers have in their leaders … It is an institutional article of faith. Level zero doctrine. On a micro and macro scale, soldiers never let you down. [Civilian] Society runs on regulation, the Army principally runs on trust.[85]

Army doctrine stresses that trust is crucial to combat effectiveness. According to this logic, if the Military Covenant is not upheld, military effectiveness is jeopardised.

Relational contract theory – in which the 'letter of the law' of a legal contract is a small part of the transaction – is mirrored by the development of psychological contract theory, a concept used to analyse the relationship between employer and employee in the civilian sphere. Kay suggests that in any workplace, the importance of relational contracts cannot be overlooked: 'working to rule' – the strict observance of the legal contract – is often used as a form of industrial action. 'The formal legal contract is not the substantive agreement'.[86] Similarly, the psychological contract concerns the relationship between employee and employer that goes beyond the clauses of the formal employment contract. The term was first formally used by Argyris in the early 1960s, then developed by Levinson et al. who argue that the individual employee and the organisation both hold

80 ADP, 2005, paragraph 715.

81 ADP, 2008a, paragraph 27.

82 Ibid., Introduction.

83 Ibid., paragraph 21.

84 Joseph J. Collins and T.O. Jacobs, Trust in the Profession of Arms, in Snider and Watkins and (ed.) Matthews, 2002.

85 Andrews, interview.

86 Kay, 1993, p. 55.

expectations of the other, which are not incorporated into the terms and conditions of the employment contract. Schein defined the psychological contract as 'a set of unwritten, reciprocal expectations between an employee and an organisation'. Argyris stated: 'Expectations such as these are not written into any formal agreement between employee and organisation, yet they operate powerfully as a determinant of behaviour'. [87] The concept was given further impetus after 1989, when Denise Rousseau's seminal paper offered a new interpretation of the contract by focusing on the individual employee's perceptions of the obligations owed towards her or him and their possible breach or violation.[88] Subsequent literature has tended similarly to focus on the employee's understanding of explicit and implicit promises regarding his or her contribution, rather than emphasising the two-way exchange between employer and employee. Guest acknowledges the difficulties in analysing a construct that concerns subjective 'perceptions, expectations, beliefs, promises and obligations'; however, he suggests the psychological contract is 'the key indicator of the contemporary employment relationship'.[89]

Psychological contract theory adds another dimension to the analysis of the Military Covenant. Like any agreement, the Covenant can be jeopardised by the non-fulfilment of its obligations, but it can also be put at risk by any breach of the psychological contract between the individual soldier (i.e. the employees) and their employers or their agents. The unique nature of military service means that employers could be perceived by soldiers to include the Army as an institution, the regiment, the chain of command, the Ministry of Defence, or, indeed, the nation, whether civilian society, policy-makers, or both. In her Covenant analysis, Tipping acknowledges the psychological contract: 'something present in all employment relationships, alongside the formal, written contract'.[90] As an organisation – 'a social arrangement for achieving controlled performance in pursuit of collective goals' – the Army cannot enter into a psychological contract. However, as Briner and Conway suggest, employees often anthropomorphise organisations, which provide them with a focus for their loyalty, commitment and affection.[91] In his evidence to the Armed Forces Select Committee, Rubin asks for consideration to be given to the psychological contract:

> … That unwritten 'contract' containing a package of expectations regarding terms and conditions of service which conferred on members of the Forces a sense of psychological security derived from principles of mutuality.[92]

87 Argyris, 1960; Schein, 1980; cited by Briner and Conway, 2005, p. 11.
88 Denise M. Rousseau, 1989, pp. 121–139.
89 Guest and Conway, 2002, pp. 22–38.
90 Tipping, 2008, pp. 12–15.
91 Briner and Conway, 2005, p. 127.
92 Rubin, HC 154-II, Appendix 7, 2000–01.

He suggests that included in a soldier's psychological contract might be providing covering fire during operations or providing anti-malaria tablets at the earliest moment before deployment.

The psychological contract, like the Military Covenant, depends upon the moral concepts of trust and reciprocity. However, any discussion of it comes with the caveat that there is a certain lack of clarity around exactly what it is. Despite the psychological contract being an example of reciprocity and exchange, studies appear unbalanced and one sided: much research and academic analysis focuses on management that usually fails to live up to its side of the bargain. While protean, subjective and implicit, the psychological contract shapes the day-to-day relations between employer and employee in the workplace. In their study of recent MBA graduates, Robinson and Rousseau found 55 per cent of the sample said the psychological contract had been broken by their employers in their first two years of employment, with psychological contract breach being the norm rather than the exception. They argue that the perception of a breach came to be negatively related to an employee's trust in their employer, job satisfaction and intentions to remain in the firm. Studies by Coyle-Shapiro and Kessler confirm the link between the psychological contract and organisational citizenship behaviour, with the fulfilment of the contract having a positive effect on employees' conduct.[93]

Little research in Europe has been undertaken in connection with the military and the psychological contract. Van der Ven argues this situation should change: 'Implementing the concept into military settings will help explain why recruits leave during initial training, satisfaction and commitment levels drop'.[94] Although comparatively neglected by European militaries, the psychological contract has been influential on those who drew up the United States' Social Compact. Like the Military Covenant, the Compact is not a legal agreement, but provides a set of mutual expectations for all parties to it.

> Our Social Compact is not just a promise, it is an exchange ... We are not talking about legalistic, documented agreements. We are talking about perception, interpretation and sense-making.[95]

However, within Social Compact research, a distinction is drawn between social and psychological contracts, with social contracts defined as 'public-shared views about what the partners in employment relationships should give and get'.[96] According to this interpretation, the Military Covenant is a social contract between Army, nation, the individual soldier and their families, while the private psychological contract – 'residing in the mind of workers and guiding their

93 Coyle Shapiro and Kessler, 2000, pp. 903–930.

94 van der Ven, 2007.

95 MacDermid et al,, 2001.

96 Ibid., p. B8.

behaviour, but not necessarily observable to others'[97] – would primarily concern the individual soldier.

In Britain, Continuous Attitude Surveys have shaped personnel strategy, first in the Army since the mid-1990s, then across the Armed Forces since 2007. The Surveys are a gauge of morale within the Forces. Along with the matters that are readily quantifiable, such as pay and accommodation, the Surveys pose questions about affective matters, such as feeling valued, that cannot be included in the Terms and Conditions of Service. While there has been little academic examination of the psychological contract and the military in Britain, it has been known as the 'conceptual contract' and taught on the Army's Junior Officer Leadership Programme. On the Army Rumour Service website, one participant shows his understanding of the difference between a legally-based employment contract and the psychological contract: (*sic*, throughout)

> Now as we all know, civvys get a contract with everything written down, pay, working hours, what they're expected to do and so forth. We get our Terms of Service which specify how long you're employed for, but that's about it.
>
> The unspoken rule is that we are expected to do everything, including putting ourselves in harms way, that the Army asks of us. In return the Army does all those bits that civillian employers don't do, provide housing, food, fly you back for compassionate cases and so on. With a fair few recent casualties we seem to be keeping our end of the bargain but are the Army?
>
> I'm interested to know what other ARRSErs think. Is the Chain of Command keeping their end of the deal in return for all that we do? From what I've seen in my time we're really good at things like training and compassionate cases but pretty poor at pay, admin and housing matters. What could the CofC improve upon?[98]

In this example, the employer is perceived to be both the Army and the chain of command, while the principle of reciprocity is inherent within the 'unspoken rule'. There is 'not much written down' – which is where problems can arise, with the subjective, perceptions-based psychological contract liable to be jeopardised if the impression takes hold that the chain of command is not 'keeping up their end of the deal in return for all that we do'. Even in the absence of legal sanctions, the norm of reciprocity ensures that on an individual level, obligations are discharged. However, echoing Durkheim, Gouldner suggests that on a macro level, inequalities relating in particular to goods and services are a barrier to truly functional reciprocity: 'power disparities allow one party to coerce the other'

97 Ibid., p. B2.

98 Cheerful Subbie, ARRSE, 06/07/06.

which can lead to a sense of exploitation and injustice.[99] Successive governments have 'constantly tested the goodwill of the Armed Forces' according to one senior officer, a view shared by the majority of his colleagues interviewed.[100] As a party to the Military Covenant, the Government – like the chain of command – should ensure it takes continual steps to uphold the psychological contract with soldiers to ensure the Covenant is kept.

The Military Covenant, based on the moral principle of reciprocity, came into collision with legal assumptions about the duty of care during the Deepcut inquiry. One of the most controversial episodes involving the Army in recent years, Deepcut blighted the Army's reputation, undermining its claims to be an employer mindful of its legal and moral duty of care to trainees. It also highlighted to some within the Army that, although the Moral Component of Army doctrine was codified, the standards of conduct it espoused had remained an aspiration rather than an implemented reality.

Deepcut brought the Military Covenant to the attention of policy-makers, heralding its future migration. The Select Committee's 2005 Inquiry, which examined the Army's duty of care from a moral as well as a legal standpoint, quoted briefly from *Soldiering*, acknowledging 'The "Moral Component" of the Army's Military Covenant'.[101] Conceding that the initial training given to some 20,000 recruits each year provided the foundation for the operational effectiveness of Britain's soldiers, the Committee was damning about the failure to fulfil the moral duty of care to young recruits:

> For far too long in the past the Armed Forces, and the Army in particular, failed to grasp the nettle of duty of care. Arguments about the level of resources available and the need to divert resources to the front-line should not have been used to mask the Armed Forces' failure to tackle these issues in that period.[102]

Concluding that 'bullying exists in the Armed Forces and is under-reported',[103] the Committee castigated the MoD for focusing on the legal dimension to its duty of care to recruits, at the expense of the moral aspect. This echoes the contract-covenant dichotomy and indicates, as this chapter has identified, the inadequacy of the purely legal dimension to good relations within any workplace. Like the Defence Committee, the Blake Review into Deepcut considered the Armed Forces' duty of care to recruits from the moral as well as the legal perspective. Blake also focused on the 'practical content' of the Military Covenant.[104] Underlining the

99 Ibid.
100 Andrews, interview.
101 HC 63-I, 2004–05, paragraph 27.
102 Ibid., paragraph 375.
103 Ibid., paragraph 274.
104 Blake, 2006, p. 391.

Covenant's reciprocity of obligation and calling on the Army to meet its side of the compact, Blake stated:

> In the opinion of the Review, the spirit of self-sacrifice referred to in the Military Covenant should not mean that soldiers have to put up with sub-standard accommodation and sanitary and washing facilities because there are other pressing demands on the defence budget. As a people-based organisation, the Army must put the welfare of its people first.[105]

The Review recommended that families were given reasonable assistance with respect to legal costs arising from inquests or fatal accident inquiries: 'In the opinion of the Review there is a good case for the Military Covenant to be interpreted' as requiring it.[106] Although Deepcut reflected the Covenant between the Army and the individual soldier had been breached, the inquiries engendered by it also gave the Covenant greater recognition in both the military and civilian spheres. In addition, Blake's interpretation of the Military Covenant heralded the concept's future migration, not only from the military to the civilian sphere, but also from the moral to the legal sphere.[107]

Conclusion

A moral agreement, the Military Covenant codified a tacit but historic understanding between soldiers, Army and nation. By making explicit what previously had been implicit, following its migration, the Covenant has compelled soldier, Army and nation to reflect upon its terms. In considering the Military Covenant in the context of other forms of contract, it is clear that there is a gulf between the restrictions and limitations of legal contract and the more expansive flexibility of an agreement such as the Covenant. However, both must be honoured. Failure to do so could adversely affect the individual soldier's psychological contract, with the risk to organisational citizenship behaviour and retention.

As the 2005 Military Covenant itself warns, goodwill and then trust will evaporate if it is not upheld. Historically, British soldiers were treated by the nation with distrust: the legal constraints upon them today that govern their conduct imply that, despite favourable opinion polls, the nation's trust should not be taken for granted. Following the Covenant's migration, many civilians seized upon the concept, among them policy-makers, the media and the third sector. As the following chapters explore, this has had many positive effects, not least in raising the profile of people issues. However, by making explicit an agreement that was instinctively understood, with subsequent claims that it has been undermined,

105 Ibid., paragraph, 12.58.
106 Ibid., paragraph, 12.110.
107 Chapter 7.

the Military Covenant may have actually been instrumental in denting the trust between soldier and Army on one side and government on the other, not least because it raised hopes and expectations among soldiers that could not be realised.

The Military Covenant was deliberately cast as a covenant. Historic, if not ancient, covenantal agreements are palimpsests for the Covenant, lending it their moral authority. If the Military Covenant is akin to the one-sided grant like the Biblical covenants, then soldiers, whose Covenantal commitment is to serve, will continue to do so, despite the nation or the Army not upholding the Covenant. This would not only be taking advantage of the military's goodwill, but when viewed in the covenantal context, is a breach of faith.

Senior commanders involved in developing the Moral Component in the 1990s are clear that the Military Covenant is not, and was never intended to be a contract in any sort of statutory or legal sense. Roberts reinforces the spiritual aura surrounding the Military Covenant by likening it to a creed.

> It was designed as a creed, it is a creed and I think creeds are better, because they are seen as aspirational rather than dogmatic. It is a creed to believe in but not to go to law about.[108]

Not only is the Military Covenant a 'covenant' not a 'contract', but it is a very specific form of trust-based moral agreement that has at its core a transcendent aspect. A legally-based covenant in the context of the Military Covenant is a contradiction in terms, reducing the unique nature of military service to the transactional.

108 Roberts, interview.

Chapter 4
The Military Covenant and the Nation: The Public

Introduction

Giving the BBC's 2006 Dimbleby lecture, the former Chief of the General Staff, General Sir Mike Jackson, referred to a reciprocal 'contract' between the nation and the Armed Forces. 'It is our soldiers who pay the cost in blood; the nation must therefore pay the cost in treasure'. As well as calling for more funding, Jackson spoke of the cultural divide between the military and civilian spheres, describing a civilian 'failure, even an unwillingness, to understand the fundamental nature of the ethos of soldiering'.[1] The following year, Jackson's successor, Sir Richard Dannatt, suggested that, because the public either opposed the Iraq intervention or misunderstood the Afghanistan mission, it was indifferent to soldiers' achievements. He implied this could have an impact on military effectiveness:

> As our operational commitments have become more intense, so has the need for support from the nation … We are in danger of sapping our volunteer army's willingness to serve in such an atmosphere again.[2]

The CGS called for Britain to emulate the example being set across the Atlantic: 'In America, appreciation for the Armed Forces is outstanding and, frankly, I would like to be able to mirror some of that here'.[3] His appeal to the public to give soldiers greater moral support was made in the context of the Military Covenant:

> … The Covenant says that we do what we do in the nation's name, that's the way a democracy works – and so soldiers do not ask why; but they do ask for respect and honour for doing what they have been sent to do, which they do with courage and professionalism.[4]

While both Chiefs considered that the civil–military relationship was under stress, their approaches were different, reflected by Dannatt going public with his concerns, and criticism of the government, while in post.

---

1 Jackson, BBC, 6 December 2006.
2 Dannatt, IISS, 21 September 2007.
3 Ibid.
4 Ibid.

The fundamental principle of the Military Covenant is that soldiers' service and sacrifice will be reciprocated by the nation's support – both material and moral. However, from late 2006, senior commanders, policy-makers and academics identified that an exceptional rift had developed between Forces personnel, particularly soldiers, and the public. The public's apparent estrangement from the military was confirmed by media reports about discrimination against – or abuse of – Service personnel and its indifference, not only to Operations Telic and Herrick, but also to the missions' impact on the welfare of serving personnel, their families and veterans. However, this chapter contends that the civil–military 'gulf' was a matter of perception rather than reality: support for the military was latent rather than non-existent. When given a lead, not least by General Dannatt, the public began to express its appreciation for Service personnel – once they had been decoupled from the missions in which they were involved. The public's unprecedented moral support, in turn, galvanised policy-makers into re-visiting 'people' issues.[5]

This chapter examines how the public came to uphold the Military Covenant and re-set the civil–military relationship. The Covenant's migration from the military sphere is charted and the concept's catalytic effect on mustering moral support for the military is assessed. Military doctrine states that the Covenant's 'greatest manifestation' is 'in the annual commemoration of Armistice Day, when the Nation keeps Covenant with those who have made the ultimate sacrifice, giving their lives in action.[6] After 2007, quite spontaneously, the townspeople of Wootton Bassett kept covenant with the fallen, the impact of which is assessed. Help for Heroes is studied, along with the support for other Forces' charities, as conduits for the public to honour the Military Covenant.

'Tommy, Go Away'?

The Military Covenant introduces a new dimension to contemporary civil–military relations analysis in Britain. With the exception of two periods of conscription in the twentieth century, a civil–military 'gap' has been the norm: as Strachan suggests, the division between the Armed Forces and the nation is 'wholly uncontroversial'.[7] Historically, soldiers have generally been regarded as a part of, yet apart from, civilian society, which frequently withheld its moral and material support, especially during times of peace. The description of the Navy as the 'senior' Service was apt in terms of funding and public approbation until the First World War. British soldiers only came to be appreciated by the public in the decades after the Crimean War when they were identified as agents of beneficent Empire – and were overseas. In *Tommy*, Kipling conveys the public's oscillating

5 Chapter 5.

6 ADP, 2000a, paragraph 103. Army Doctrine Publications. Ministry of Defence. © Crown copyright 2014.

7 Strachan, 2003, pp. 43–63.

attitudes towards soldiers, which veered from overt hostility during peacetime to adulation 'when the guns begin to shoot'. The Army's absence made the British public's heart grow fonder: the civil–military gap of the late Victorian era was ideally an ocean or two. Even during the highpoint of Empire, Britain was not militarised to the same extent as other European states. Shaw highlights the contrast between the country's military ambitions and military activism, and the relatively small number of Forces' personnel compared with Continental powers.[8] Although both World Wars – a time of 'democratic militarism'[9] – enmeshed the Forces more deeply in the national narrative, defence has been in 'managed decline' since 1945. Since the 1960s and the end of conscription, the Armed Forces have remained integral to British national life but, generally, at the periphery of it, a situation best symbolised by the Army's presence in Northern Ireland. IRA terrorism drove soldiers 'into civilian clothes and behind barbed wire' for almost three decades and reduced the public visibility of soldiers on the British mainland.[10]

By 2006, a rift was perceived to have emerged in the civil–military relationship that went far beyond the usual gap between the two spheres. This civil–military gulf was the result of Iraq, an issue that blighted British public life. Service personnel appeared to be scapegoats for the intervention, for which, as Kerr observes, they bore the 'burden of illegitimacy'.[11] Opponents of the intervention – on the eve of which the biggest demonstration in British history had taken place – would never be reconciled, while the initially favourable polls caused by the inevitable 'rally round the flag' effect and a 'Baghdad bounce' rapidly reversed. A national security rationale was unconvincing in the absence of finding WMD in Iraq and the July 2005 London bombings. By March 2007, a BBC/ICM survey found 60 per cent of those polled thought the intervention was a mistake.[12] Similarly, from 2006, there was little public engagement with the mission in Afghanistan. In October, 64 per cent reported that there was no clear strategy, 81 per cent said the British forces were overstretched, only 3 per cent expressed any confidence in the government's handling of the situation, while 77 per cent were proud of the way that British Forces were performing their duties.[13] By 2009, more than 60 per cent said the war in Afghanistan was unwinnable and that all British forces should be withdrawn, while 42 per cent did not understand the purpose of the British mission.[14] In October 2006, a *Guardian*/ICM poll had found that 45 per cent wanted an immediate withdrawal of troops from Iraq, while a YouGov-Telegraph survey reported that the majority believed there was a realistic chance of British Forces being defeated in Afghanistan (58 per cent) and

8 Shaw, 1991, pp. 109–162.

9 Ibid.

10 Dannatt, interview.

11 Kerr, 2008, pp. 401–419.

12 BBC/ICM Iraq Survey, 2007.

13 YouGov/*Daily Telegraph*, October 2006.

14 BBC Politics Show/ComRes, November 2009.

Iraq (66 per cent).[15] It is unsurprising that Chiefs Jackson and Dannatt found the public's attitude wanting.

The Military Covenant codifies that, in exchange for their service, soldiers must expect 'to be valued and respected as individuals'. In November 2005, a *Sunday Times* front-page story concerned public indifference to frontline soldiers, who believed 'that the MoD is not supporting them and nobody in the UK cares about what is happening in Iraq'.[16] From 2006, a spate of incidents suggested that soldiers inspired not just apathy but antipathy among the public. Such incidents were rare, but each one created a media storm. The most notorious were alleged attacks on the wounded at Selly Oak hospital, home of the Royal Centre for Defence Medicine.[17] Hospital representatives rebutted the reports, describing them as 'inaccurate, unsubstantiated and ill-informed'.[18] During questioning of the Hospital's Chief Executive by the Defence Select Committee, repeated references were made to 'urban myths', while four Birmingham MPs implied that unsubstantiated stories were being repeated by some MPs and senior commanders as a pretext for attacking the decision to close military hospitals.[19] Although the Committee condemned much of the newspaper coverage as 'inaccurate and irresponsible', Colonel Stuart Tootal, formerly of 3 Para, who visited the wounded in Selly Oak, contrasted the 'excellent' standard of healthcare soldiers received in the field with that they received back in Britain; 'a disgrace'.[20] In one of the first references to the Covenant by a civilian, in June 2007 the Committee chairman asked whether mixed civilian–military wards were a 'breach of the Military Covenant'.[21] This presaged how the concept would encompass both the material support and moral support given to soldiers – and would be perceived in negative terms.

The Covenant's migration to the civilian sphere was an Army initiative to narrow the civil–military gulf. As CGS, Dannatt saw part of his job as encouraging the public better to 'identify and engage with the military'.[22] From 2007, policy-makers and academics were similarly concerned about the apparent civil–military rift. In 2008, Prime Minister Gordon Brown said it was 'vital' for Service personnel to know that the whole of Britain 'understands and appreciates the work they do in our name'.[23] The Government-backed 2008 National Recognition Report (The Davies Report) saw this in terms of improving understanding, support, visibility

---

15 Glover et al., *The Guardian*, 24 October 2006; YouGov/*Daily Telegraph*, October 2006.

16 Smith, *Sunday Times*, 6 November 2005, p. 1.

17 Muslim Accosts Injured Para in Hospital, *Daily Telegraph*, 2 October 2006; Muslim Women Abuse Soldiers at Troops' Hospital, *Mail on Sunday*, 10 June 2007.

18 UHB NHS Foundation Trust, 2006–07, p. 11.

19 Lynne Jones MP et al., Memorandum, HC 327, 2007–08, ev 112.

20 Tootal, interview.

21 James Arbuthnot MP, HC 327, 2007–08, ev 24.

22 Dannatt, interview.

23 MoD, 2008a, Introduction.

and contact.[24] It stated that 'understanding of the military and recognition of their role' determines the taxpayers' willingness to finance the Forces, as well as the climate in which they recruit.[25] In 2007, Edmunds and Forster focused on the gulf in 'understanding', particularly in connection with operations:

> There needs to be a wider debate and much greater public understanding about the type of complex missions that our armed forces fight, the contribution of British armed forces to our security and the risks they face on our behalf.[26]

According to them, the Covenant was 'broken almost beyond repair'.[27]

This study suggests the public had neglected rather than broken the Military Covenant. Its support for soldiers was latent, not non-existent: given the opportunity, it was eager to express its appreciation. Assertions about negative attitudes towards the military are at odds with successive surveys of public opinion. According to polls published after 2003, approval ratings for the Armed Forces were something that politicians could only wish for. In late 2008, Ipsos MORI found that 81 per cent had a favourable view of the Forces, the highest level on record: a 2003 MoD survey found support at 74 per cent.[28] Conversely, Tony Blair's net approval rating in connection with Iraq plunged to minus 60 by May 2007.[29] Polls undertaken for the MoD since February 2000 were consistently positive, with more than two-thirds saying that the Armed Forces had the highest professional standards. A 2006 survey found that 87 per cent agreed that 'the British Armed Forces are among the best in the world'.[30] In 2009, a National Army Museum/ICM study found that, while 60 per cent disagreed with the deployment of troops to Iraq and 53 per cent were against the deployment to Afghanistan, 64 per cent would support their child joining the Army.[31]

The public's implicit but staunch moral support for Forces' personnel did not equate to backing for the missions in Iraq and Afghanistan. In January 2007, in the speech that referred to the Covenant, Tony Blair stated:

> They [Service personnel] want public opinion not just behind them but behind their mission. They want the 'people back home' to understand their value not just their courage.

24 Ibid., pp. 6–26.
25 Ibid., p. 4.
26 Edmunds and Forster, 2007, p. 13.
27 Ibid.
28 Bailey, *Daily Telegraph*, 1 November 2008.
29 Ipsos MORI, War with Iraq: Trends 2002–07.
30 Ipsos MORI, January 2006, cited by Seldon, 2009, p. 114.
31 National Army Museum/ICM, 2009.

With even the CGS unconvinced about the Iraq intervention, this implies a certain amount of wishful thinking by the Prime Minister.[32] Although the public came to express its appreciation for soldiers, narrowing this aspect of the civil–military gulf, the divide in terms of support for the missions grew steadily wider.

The civil–military relationship in Britain today is contradictory. In 2006, while Chiefs Jackson and Dannatt perceived a divergence between the civil and military spheres, according to that year's Continuous Attitude Survey, the overwhelming majority of soldiers themselves (77 per cent) were proud to be in the Army.[33] Although the public did not necessarily understand either the unique nature of military Service, the Service ethos or the challenges military personnel faced in Basra or Helmand, they trusted in the military's professionalism to do the job well. In addition, while polls reflect moral support for the Armed Forces, this would not necessarily translate into material support. According to Ipsos MORI, in 2006, while 33 per cent of public said that the UK spent too little on defence, spending on other areas – whether the NHS, education, the police and overseas aid – was considered a higher priority.[34] During the week Britain withdrew from Iraq in 2009, only 9 per cent of the public viewed defence as the most important issue facing the nation, the lowest level since 9/11.[35]

The perceptions of a civil–military gulf were fuelled by the media and policy-makers. While opinion polls are a reliable gauge of public opinion, they do not quite convey attitude, 'a predisposition to respond in a favourable or unfavourable manner to a particular object or class of objects'.[36] Attitudes are affective as well as cognitive and can determine behaviour. Although the unpopularity of Iraq and the lack of public appreciation for soldiers were linked, both former CDS General Lord Guthrie and Tootal suggest antipathy to soldiers as scapegoats for the mission and to the mission itself was exaggerated, pointing out there was nothing in Britain on a scale of the Vietnam-era anti-war demonstrations. As Tootal observes, 'soldiers haven't been spat at in the street'.[37] Although the Iraq intervention remained unpopular, the level of active protest against it petered out. Equally, the few reported incidents of discrimination against soldiers, even when substantiated, were the result of actions by individual employees who appear over-zealous rather than prompted by anti-military or anti-war sentiments. Refusing to give a cheaper fare to a soldier travelling without a warrant or barring uniformed Army officers from a department store surely indicates individual jobsworthiness rather than hostility to the military in the nation at large.[38] Refusing to sell alcohol

32 Dannatt interview, Sarah Sands, 12 October 2006.

33 MoD, Armed Forces Continuous Attitude Survey, March–July 2006.

34 Ipsos MORI, in Edmund and Forster, 2007, p. 35.

35 Ipsos MORI, Issues Index, April 2009.

36 Oskamp and Schultz, 2005, p. 18.

37 Tootal, interview.

38 Harrods Bans Soldiers on Poppy Day, *Daily Mail*, 18 November 2006.

to uniformed soldiers in a pub or at a petrol station appears to have been based on employees' interpretation of Army regulations.[39]

The Davies Report stated that the foundation of support among the nation for the Armed Forces had 'eroded over recent years'.[40] This statement was based on little quantitative evidence. Of the four examples of discrimination against Armed Forces personnel cited, two were the result of pre-emptive action by senior officers in early 2007 who, anticipating hostility, banned personnel wearing uniforms in public. In the first instance, the ban related to personnel from RAF Wittering visiting Peterborough; in the second, soldiers arriving at Birmingham Airport from Afghanistan. Both bans were the result of erroneous assumptions about the public's attitude, rather than the public's attitude per se. Indeed, a subsequent poll found that 90 per cent of those questioned said that Armed Forces personnel should be able to wear their uniforms in public.[41] However, initial media reports fed a growing perception, certainly among Forces personnel, that the public mood was anti-military. Davies states: 'These incidents were very frequently and spontaneously referred to by the serving servicemen and servicewomen we met'. Davies appears not to have investigated the incidents in any depth and failed to note that, far from being proof of anti-military sentiment, each one actually provoked a public response that was overwhelmingly supportive of the Armed Forces. A recent poll found that while 21 per cent of soldiers have experienced strangers shouting abuse at them when in uniform in the UK, 56 per cent have been thanked.[42] Press reports about the negative attitude of local residents to Headley Court – based more on Nimbyism than prejudice against soldiers – raised awareness of the institution and garnered a 33,000 signature Downing Street petition signature in its favour.[43]

Even if conclusions about national attitudes towards the military should not be drawn from a small number of incidents, in late 2006, CGS Dannatt considered the value and respect demanded of civilian society by the Military Covenant was absent, an opinion shared by interviewees. However, this study suggests that rather than being antipathetic to the military, the public was apathetic, withholding tangible expressions of support. Public appreciation of military professionalism was not helped by the media ban on reporting from the frontline in Afghanistan. Dannatt says:

> There was an embargo effectively, through the Ministry of Defence press office, on the media from reporting war-like activities ... The effect of that on soldiers out there at the time was to make them very angry, and I, quite rightly had my ear

39 Asian Petrol Station Worker 'Refused to Serve Afghanistan War Veteran Because He Was in Uniform', *Daily Mail*, 24 October 2007.

40 MoD, 2008a (Chapter 6).

41 ICM/*Sunday Telegraph* 12/13 March 2008.

42 Ashcroft, *Armed Forces and Society*, 2012.

43 Gillan, *The Guardian*, 21 July 2007.

> bent in no uncertain terms in Autumn 2006: 'you know we're out here, fighting and dying, but they [the public] don't know anything about it'.[44]

Although outside the scope of this study, this initial media ban seems curious. Former Guards Officer Adam Holloway MP argues that media attitudes towards the military have changed since he was an ITN correspondent in Bosnia: since then the press corps has been dependent on the military in theatre 'for logistics, protection, medicine, food, fuel and booze'. He says:

> As a result the media has much more come round to 'the boys'. There have been misgivings among the public about what we are doing in these places [Iraq and Afghanistan], but sympathy for the guys: that has been helped by having a press that has much more understanding now than in the past.[45]

The practice of 'embedding' journalists has promoted greater military-media contact and empathy. James Fergusson considers embeds 'suck' and has never been on one:

> Journalists for big news organisations depend on future embeds for a living and would be silly to bite the hand that feeds them. There are plenty of examples of journos being busted off embeds at the last minute (or even during) because they have been critical of the military in the past.[46]

Accompanying soldiers into combat has an emotional impact, which as Fergusson suggests, tends to impair objectivity. Cowper-Coles observes, in those circumstances, it is difficult to write negative things about soldiers or to question the point of the mission: 'That is of course one reason why defence ministries are eager to embed journalists'.[47] With home-front hearts and minds needing to be won, the initial media ban appears counter-productive.

From 2007, the public began to express its hitherto implicit support for soldiers, narrowing the civil–military gulf, which had been expressed in terms of a fractured Covenant. Brigadier (retired) Ed Butler observes there was a 'quantum shift in the nation's support for the Armed Forces'.[48] Before deployment, he had raised his concerns about the lack of recognition being given to soldiers with senior politicians and commanders. He believes public indifference was due to the Government not clearly communicating the ambitions and objectives of the mission in Afghanistan, a result of 'the Government not knowing what they wanted themselves'. Tootal, whose soldiers were not given a homecoming parade

---

44 Dannatt, interview.

45 Holloway, interview.

46 Fergusson (email exchange).

47 Cowper-Coles, 2011, p. 98.

48 Butler, telephone interview.

in 2006 and were often refused entry to bars and nightclubs in Colchester, said there has been a 'large change in the public mood' with 'far greater recognition of what soldiers go through'.[49] A former MoD official observes there was 'a huge improvement in public recognition'.[50] Officer L agrees and asks rhetorically: 'What triggered it?'[51]

Migration

The Military Covenant was a concept invented by the Army that began its migration to the civilian sphere six years after it was published. This study contends that in October 2006 General Dannatt started this process, which was given further impetus by Prime Minister Blair in January 2007, in a speech aboard *HMS Albion*. Pivotal to the Covenant's success was the launch of first, the Royal British Legion's *Honour the Covenant* campaign and, almost concurrently, of Help for Heroes. The public's support for soldiers, mobilised by the media and the third sector, in turn influenced politicians who, from September 2007, scrambled to catch up, their attention finally focusing on the 'people issues' affecting soldiers and the wider Forces' community.[52] However, it was the broadsheets, the Legion and Opposition politicians who conveyed support for the military in the Covenant context. Help for Heroes was more populist, urging the public to 'back our boys and girls'. Encapsulating the charity's approach – and explaining its success – is its poster depicting two groups: on one side are Servicemen, some wounded; on the other, England footballers. Both are lined up above the slogan '*Different Duties. Same Pride*'.[53]

The Military Covenant's successful entrenchment in the civilian sphere was far from guaranteed. The dynamics of social trends whether relating to products, ideas or norms have been examined by economists, epidemiologists and sociologists, who bring different approaches to 'macro-behaviour'[54] and 'diffusion'.[55] Conceptual frameworks, often involving mathematical models, can explain *ex post facto* the process of spontaneous and unplanned success, but they do not explain why certain new concepts or consumer goods succeed. Rogers suggests that many innovations either do not succeed (e.g. Betamax) or do so slowly.[56] Whether involving the concrete or the abstract, what succeeds and becomes entrenched appears to be driven not only by rational self-interest, but also by the

49 Tootal, interview.
50 Non-attributable, interview.
51 Senior Commander, interview.
52 Chapter 6.
53 England Footballers Foundation, 2011.
54 Schelling, 2006.
55 Rogers, 2003.
56 Ibid., pp. 7–10.

human need to belong, and what Olsen describes as 'social incentives' such as prestige.[57] Innovations can be adopted on an individual or organisational level and can be disseminated by horizontal social networks and by the mass media. They can reach critical mass – 'the rate at which diffusion becomes self-generating and non-reversible'[58] – seen on a global scale with the adoption of email; or a 'tipping point' – 'that magic moment when ideas, trends and social behaviours cross a threshold, tip and spread like wildfire' – examined by Gladwell, based on work by Schelling, who analysed 'tipping' in the context of racial segregation within neighbourhoods.[59] Similarly, contingent behaviour – 'behaviour that depends on what others are doing'[60] – can account for the 'bandwagon effect' identified by Liebenstein or the 'threshold' explored by Granovetter. The dynamics behind Hardin's 'tragedy of the commons'[61] and short-lived fads such as the Go-Go Hamster, the under-eights' must-have toy of 2009, can be explained. However, as the toy's manufacturer stated: 'People ask us how does this happen. The answer is that if we knew that, we'd have a hit every week'.[62] While the dynamics of success can be charted, success cannot be predicted.

The momentum that led to the Military Covenant's migration and entrenchment is, with hindsight, an example of what Schelling suggests is a collective action that almost has the appearance of being organised. Dannatt's advocacy of the concept reflects how the action of individuals can have an aggregate effect, regardless of their motive: Prime Minister Blair's acknowledgement of the Covenant, which further legitimised it, concerned the defence of his government's record, rather than support for soldiers. Dannatt himself sees his September 2007 speech as key:

> I think that was the start of it. When you throw a rock into the pond, the ripples die away. Well, those ripples just got bigger. Other people picked it up and on it's gone.[63]

The speech gave further impetus to a process of low-key migration that had been underway for almost 12 months, as senior officers placed 'people issues' in the Covenant context when discussing them in the media.[64] As Dannatt suggests, this was an example of the Army's ethos of Mission Command: 'What I was doing was giving a lead and others picked it up. I'd be very disappointed if they hadn't'.[65] Rogers' analysis of the innovation process within organisations suggests

57 Olsen, 1971, p. 60.
58 Markus, 1987, pp. 491–511.
59 Schelling, pp. 137–166.
60 Liebenstein, 1950, pp. 183–207.
61 Hardin, 1968, pp. 1243–1248.
62 Russell Hornsby, in Birchall, *Financial Times*, 8 November 2009, p. 1.
63 Dannatt, interview.
64 E.g. Viggers in Evans, 4 January 2007, p. 8.
65 Dannatt, interview.

that 'innovation champions' can play a crucial role in boosting a new concept.[66] Although the Covenant was not new, its deployment to convey the Army's concerns to an external audience was.

With the launch of Help for Heroes, civilians became Military Covenant champions – usually unwittingly. The charity's co-founder, Bryn Parry also uses the ripple analogy to explain how mass support for it was generated following a fundraiser at Twickenham:

> With the rugby match, 52,000 people left that stadium all thinking what could they do to help support the men and women of the Armed Forces. It was very powerful. I always refer to it as being the pebble in the pond. If you drop the pebble into community ponds, by hitting one contact or communicator, you spread the word to the whole network.[67]

Parry captures the process of horizontal transmission that contributed to Help for Heroes' success. This was augmented by a 'bandwagon effect' among celebrities who were eager to support the charity, which in turn engendered media coverage. Unlike the Legion, Help for Heroes did not frame support for soldiers in the Covenant context: however, by promoting the welfare cause, the charity unwittingly promoted the concept. The difference between the two charities is that Help for Heroes was aimed at the public, while the *Honour the Covenant* campaign was more overtly political: directed primarily at policy-makers, it aimed to change policy in connection with the welfare of the Forces' community.

The Military Covenant's successful entrenchment in the civilian sphere was a contingent process, driven in part by the media. Public attention is a limited resource competed for by countless interest groups.[68] Whether the media can manufacture consent for public policy;[69] is driven by commercial interests;[70] or skew stories to attract an audience for advertisers,[71] they have limited time or space to fill. How far the traditional media lead or follow their audience or readership is a matter for debate, although one editor has argued his publication reflects its readership.[72] Conversely, in their study of the 1968 US Presidential election, McCombs and Shaw argue that the media set the agenda: their emphasis on issues corresponds with the importance of those issues to voters.[73] Indeed, McCombs goes further: the

66 Rogers, p. 414.

67 Parry, interview.

68 Barakso and Schaffner, 2006, pp. 22–44.

69 Herman and Chomsky, 1994, p. 289.

70 Davies, 2008.

71 Morris and Forgette, 2007, p. 97.

72 'My job is to edit my newspaper, to have a relationship with my readers, to reflect my readers' views and to defend their interest'. Paul Dacre, Editor, *Daily Mail* to Commons' Public Administration Committee 2004, cited by Davies, 2008, p. 370.

73 McCombs and Shaw, 1972, pp. 176–187.

media not only tell the public which issues to think about, but how to think about them.[74] However, identifying a five-stage 'issue attention cycle', Downs argues that the public manages the news by maintaining or losing interest in issues.[75] To date, the Covenant has confirmed to the first few stages of the cycle: the 'alarmed discovery' of a long-standing problem; that is, apparent indifference to Forces' personnel on active service, followed by 'euphoric enthusiasm' to find a solution.

This study has sought to determine the impact of the Military Covenant on the civilian sphere by examining national press coverage. It wanted to establish when migration began, how the concept was defined, who championed it and whether there was any link between the Covenant and coverage of welfare issues affecting Service personnel. Evidence was compiled using the Nexis UK database, with searches involving four national newspapers: *The Sun*, *The Guardian* (and the *Observer*), *Daily Telegraph* (and *Sunday Telegraph*) and *Daily Mail* (*Mail on Sunday*). With a combined daily circulation of 6.7 million in September 2007, their readership represents a broad demographic and a spectrum of political views. The two broadsheets are political polar opposites: with its 3 million-plus readers, *The Sun* is Britain's biggest-selling newspaper.

The first search sought to chart the Covenant's migration from Army doctrine. The terms 'Military Covenant' and 'All UK National Newspapers' and the timeframe 1 January 2000 – 1 October 2006 produced just two articles, both in broadsheets.[76] This suggests that after the publication of *Soldiering: The Military Covenant* in 2000, the concept was of little interest to civilians. However, a search of the period 1 October 2006 – 1 October 2007 produced 115 articles. This timeframe includes Dannatt's *Daily Mail* interview and the *Honour the Covenant* launch. Of the 115 articles, 71 were published after August 2007 when the Legion first publicised its campaign. The Legion's Director-General Chris Simpkins says: 'It was very, very quickly picked up by the media and used as shorthand for supporting the Armed Forces'.[77]

As CGS, Dannatt saw his 'primary task' as tackling people issues, to which he alerted the public.[78] To assess his success, coverage of two areas was studied; accommodation and treatment of the wounded. While the same timeframes and four newspapers were employed, various terms were searched for, including 'Army accommodation', 'Army housing', and 'soldiers' housing'. All pointed to a rise in interest after October 2006, compared with the earlier search period. For example, from 1 January 2000 – 1 October 2006, 478 articles featured 'Army Barracks': just one focused on its quality – 'slum dwellings'.[79] In the next 12 months, 11 of the 77 'Army Barracks' articles concerned its poor quality. Similarly, between

74 McCombs, 2004, p. 5.

75 Downs, 1972, pp. 38–49.

76 Evans, 3 April 2000, p. 4; Utley, 4 April 2001, p. 13.

77 Simpkins, interview.

78 Dannatt, 2010, p. 266.

79 Guthrie, cited by Smith, 26 February 2001.

1 January 2000 and 1 October 2006, using the search criterion 'Headley Court', there was one article, compared with 43 from 1 October 2006 – 1 October 2007. Although the level of sacrifice was far higher after the deployment to Helmand, with 35 soldiers being killed between March and October 2006, 19 of them in September, 118 soldiers had been killed in Iraq between 1 March 2003 and 1 October 2006.[80] These results suggest that, having framed people issues in the context of the Covenant with its inherent reciprocal moral obligations, General Dannatt was successful in raising their public profile.

Just as *Soldiering: The Military Covenant* had been written for the officer class, the Military Covenant's civilian champions tended to come from a social elite of policy-makers and opinion-formers. Among the broadsheets, *The Independent on Sunday* was initially the concept's most prominent champion. Harold Pinter was a signatory to its open letter calling on to Tony Blair to uphold the Covenant: this was followed by the paper's own Covenant campaign in early September 2007.[81] To galvanise public support for soldiers, *The Sun* explicitly separated them from the Iraq mission. Attending a homecoming parade or wearing a Help for Heroes wristband '… Shows not that you support war in Iraq. Just those having to fight it'. It continued:

> Too many people misguidedly fear that by treating our servicemen with even the slightest respect they endorse a war they disapprove of … We should feel honoured to have men and women of such courage defending our country's interests … We should not hesitate to show them respect and gratitude.[82]

The Sun said troops were 'scapegoats in a war that has divided the nation'.[83] To muster support for soldiers – a cause its readers rallied to – the tabloid was forced to play down, if not to renounce, its previous support for the intervention.

Although the public began to express its hitherto latent support for Forces personnel after 2006, few of them would have described their actions in the Covenant context. However, this support – both material and moral – generated a slipstream that propelled the concept's further migration to, and entrenchment in, the civilian sphere. Help for Heroes best reflects this slipstream effect. Weeks after the charity's launch, *The Sun* urged its readers to buy a Help for Heroes wristband. Parry says: 'They did a front page and it spread like wildfire with *The Sun's* support'.[84] From 1 October 2007 – 31 December 2010, 1521 articles in *The Sun* referred to the charity, dwarfing the 23 references to the Covenant. Similarly, the *Telegraph* (248 [69]) and the *Mail* (107[43]) favoured the charity over the

80 DataBlog, *The Guardian*, ongoing.

81 Terri Judd et al., *The Independent on Sunday*, 11 March 2007. Leader, Britain Must Honour its Commitment to Our Troops, *The Independent on Sunday*, 2 September 2007.

82 Ibid.

83 A Wake Up Call, *The Sun*, 29 October 2007.

84 Parry, interview.

Covenant in the same period, but *The Guardian* reversed the trend (27[50]). Lord Guthrie observes that *The Sun* has been 'fantastically supportive' of the Armed Forces. Starting in 2008, the 'Millies', the paper's annual Military awards, secured royal and Prime Ministerial patronage.[85] From 2007, a virtuous circle was established: Service charities gained media coverage, which encouraged the readership-public to support the charities, which because of readership-public support, garnered more media coverage. The charities' supporters might not have considered they were upholding the Covenant, but they were in fact strengthening the 'common bond of identity, loyalty and responsibility' with soldiers.

The initial media coverage of the Military Covenant was frequently inaccurate, but gave the concept a useful patina of historical authenticity. Reports claimed that 'for almost 200 years Britain has had a military covenant which lays down the mutual obligation between the forces and the nation';[86] that the Covenant was 'drafted in the 19th century and reviewed in 2005';[87] is a 'fundamental agreement between the British people and the men and women who fight on our behalf, which has existed in some form since the time of the Duke of Wellington, [and] has now been broken';[88] was 'drafted in the days of Wellington';[89] is 'an ancient pledge'.[90] In its article about the imminent launch of the Honour the Covenant campaign, *The Guardian* quoted in full the summary paragraph from *Soldiering* that defined the concept, but also said it was 'historic', a claim repeated in a subsequent report.[91] Although spurious, claims concerning the concept's historical roots bolstered its moral authority. A similar confusion surrounded definitions of the Covenant. It is a 'contract' to which soldiers agree 'when they sign on';[92] 'drawn up by Army chiefs to set out the way the nation should treat its soldiers';[93] 'the long-held principle that Britain has a duty of care towards its Armed Forces in return for their service to the nation';[94] 'the guarantee of a duty of care between Government and Armed Forces';[95] 'the duty of care the nation pledges to troops in recognition of their extraordinary sacrifices';[96] 'an unbreakable bond of loyalty and mutual obligation between the nation and its Armed Forces. It is not a legal document, but morally it is just as binding'.[97] Although this last example comes closest to the definition of

85 Guthrie, interview.

86 *Daily Mail* Comment, Shameful Failure to Care for our Forces, 12 March 2007.

87 Rawsthorne, *Daily Mail*, 19 March 2007.

88 *Daily Mail* Comment, Our Boys Betrayed, 14 September 2007.

89 Hastings, *Daily Mail*, 8 March 2008.

90 Kay, *The Sun*, 17 September 2007.

91 Gillan, *The Guardian*, 15 August 2007: White, *The Guardian*, 29 December 2007.

92 Bell, *Mail on Sunday*, 25 March 2007.

93 *Daily Mail* Comment, 16 August 2007.

94 Leake, *Daily Mail*, 9 August 2009.

95 Hastings, *Daily Mail*, 8 March 2008.

96 Ainsworth Plans Law to Let Soldiers Jump Waiting Lists, *Daily Mail*, 18 December 2009.

97 *Daily Mail* Comment, Wounded Heroes and a Question of Honour, 6 September 2010.

the Covenant as formulated in Army doctrine in 2000, its extension to the Armed Forces shows how the concept would be modified by civilians. *The Guardian*'s various definitions veer between the moral 'commitment' and 'pact'; the legal 'contract'; or quasi-legal 'duty of care'.[98] Rather than the trilateral agreement set out in military doctrine, all newspaper coverage asserts that the Covenant is a bilateral agreement between soldiers (or Forces' personnel) and government.

Thank You, Mr Atkins

Politicians and senior commanders have assumed that a lack of public understanding of the Armed Forces is automatically damaging to them. However, the two periods of conscription – 1916–19 and 1939–63 – engendered an unprecedented familiarity with the Forces, while belief in the justice of the cause being fought further interlocked the civil and military spheres. The negative perception of the First World War emerged in terms of British culture not in the 1920s or 1930s, but, as Bond suggests, in the 1960s.[99] Since then, the 'donkeys and Flanders mud'[100] view of the conflict, reinforced by popular culture, seems to be 'unassailable, despite scholarly challenges'.[101] However, the *Oh! What a Lovely War* tradition overlooks how war was viewed as a necessity. Even the conflict's most eloquent critics – Sassoon, Graves and Owen – remained 'proud of their regiments and personal achievements and deeply grateful for the unique experience of comradeship'.[102]

While 1.1 million young men were conscripted between 1949–60,[103] closing the gap in understanding, this did not necessarily correspond with appreciation, something overlooked by many analysts. According to opinion polls, in 1949, 57 per cent of the population thought National Service worthwhile and 33 per cent opposed it. By 1953, despite war in Korea, these figures were reversed.[104] The annual intake of 160,000 men disrupted industry needed by drab, austerity Britain. In 1955, a Gallup poll showed that one third of the country favoured defence cuts above any other savings.[105] As Vasquez observes, a social justice norm that ensured that National Service was indeed national, in that all classes undertook it.[106] Understanding the Armed Forces only too well, the browned-off British held them in none of the reverence with which they have been

98 Norton Taylor, *The Guardian*, 13 September 2007; Norton-Taylor, *The Guardian*, 18 July 2007; Norton-Taylor, *The Guardian*, 5 November 2007.

99 Bond, 2002.

100 Ibid.

101 Holden Reid, note.

102 Ibid. p. 31.

103 Vasquez, 2011, pp. 636–656.

104 Royle, 1986, p. 204.

105 Ibid.

106 Vasquez, 2011, pp. 636–656.

accorded since 2006. *Carry on Sergeant* says much about the national attitude towards the Army in the late 1950s. Cowper-Coles observes the portrayal of the pomposity of military figures in the comedy *Square Peg* is almost inconceivable today.[107] Familiarity did not breed contempt so much as mockery. One former senior commander observes that all national institutions whether Army, Church, Parliament or the Monarchy should preserve a certain mystique.[108] As much as many politicians and senior commanders might call for the civil–military gap in terms of understanding to be closed, what they rarely consider is that it is surely more advantageous for the Army as an institution that soldiers remain aloof from civilian society, emerging for short, sharp bursts of derring-do on foreign fields, dignified ceremonial and the collection of gallantry medals.

Charities have long been the conduits for civilian society to show its appreciation of Service personnel, veterans and their families. Britain's third sector includes some 161,000 registered charities whose income was £53.8 billion in 2010: two-thirds of this (£36.5 billion) was generated by just 1.1 per cent (1765) of charities. Among this elite group, which each raised more than £5m in 2010, are several Service-related charities, including Combat Stress.[109] With the charities frequently overlapping and duplicating the aid they offer, both a RUSI paper and a Prime Ministerial Task Force called for the sector to be rationalised.[110] According to the Confederation of Service Charities (COBSEO), while there might be scope for amalgamation, increasing cooperation can only be encouraged rather than forced.[111]

Donations to Service charities, by which the public demonstrate their tangible support for the Forces, are one metric to assess the catalytic effect of the post-migration Military Covenant. The Royal British Legion's income climbed £80.9m in 2005 to £92.2 in 2007. Two more record-breaking years followed, with £104.1m raised in 2008 and £107.3 in 2009. Accounts lodged with the Charity Commission show an upsurge in the income of the leading military charities in the five years after 2005. Combat Stress saw its income almost double from £6.73m in 2006 to £9.7m in 2008, reaching £12.55m in 2010. Similarly, SSAFA's income rose from £21.2m in 2005 to £47.24m in 2009. Donations to the Army Benevolent Fund went from £5.7m in 2006 to £12.1m in 2010.[112] That year the charity rebranded itself as the Soldier's Charity, shaking off connotations with the workhouse era. The reinvention was prompted by the impact of Help for Heroes, which, according to COBSEO forced all the sector's organisations to 'change their game and raise their game'.[113] While the Legion leads the annual Festival of Remembrance at the

107 Cowper-Coles, 2011, p. 280.

108 Former senior commander, interview.

109 Charity Commission 2010.

110 RUSI, Whither Welfare Report, 2010; Strachan, Armed Forces' Covenant Task Force, Summer 2010; Christopher Dandeker et al. 2003.

111 Bray, interview.

112 Figures from accounts filed with the Charity Commission.

113 Bray, interview.

Albert Hall, in 2008 on behalf of Help for Heroes, pop maestro Simon Cowell produced the 600,000-selling *Hero* single by X-Factor contestants – a far cry from Dame Vera Lynn. The Legion believes its income has been unaffected by Help for Heroes, but concedes there is a perception that it is 'an old man's organisation', which only works three weeks of the year during the annual Poppy Appeal.[114] Dannatt argues that 'not by design, but by circumstance' Help for Heroes 'stirred up the entire Service charity sector and provided a brilliant way for the population at large to get behind its service men and women, and their families'.[115]

Bryn Parry was unaware of the Military Covenant when he and his wife Emma founded Help for Heroes, but instinctively understood its fundamental principle of support for soldiers in exchange for their service. A former Army officer, he has experienced the scapegoating that soldiers can suffer because of an unpopular mission. He says:

> I know only too well what it's like to come back from a Northern Ireland tour and being asked to leave a pub because I was a soldier, or talking to my civilian friends, to whom I had somehow become a mass murderer or a stormtrooper. I felt I was protecting people. There was a big disconnect.[116]

Parry wanted to get across that an unpopular war must not lead to an unpopular soldier, a theme the media began to explore with the launch of *Honour the Covenant* and Help for Heroes. He believes that the 'natural default setting' in Britain is support for the Armed Forces.

> Although I wasn't aware it was called the Military Covenant, when we set up Help for Heroes we were very much working on the idea of the public wants to support the men and women of the Armed Forces, they need an outlet and a way of showing that support. Therefore we set up, with the determination not to be political or critical. We were simply wanting to help. That was our very simple message, which I think is in accord with the whole idea of the Covenant.[117]

Although a 'broken' Covenant became a shorthand term for privations suffered by an overstretched military – for which the Government was blamed – for Parry, the Covenant concerns the Forces' relationship with the public.

> The Covenant is much more about the military, how it fits within society. I think the stuff that was happening in 2007 was fact. For example, there was talk that there weren't enough helicopters. But that's purely an operational matter. The

114 Simpkins, interview,
115 Dannatt, 2010, p. 274.
116 Parry, interview.
117 Ibid.

> Military Covenant is much more to do with how we the citizens of our country support a professional Army, Navy and Air Force.[118]

Britain's fastest-ever growing charity, Help for Heroes raised more than £100m in three years after its launch. Parry says it 'deliberately went to the general public' rather than fundraising from within the Forces' community, as other military charities tend to do. Although unwittingly, Help for Heroes embodies the principles of the Military Covenant, particularly the bond of 'identity, loyalty and responsibility' it codifies. Supporters were encouraged to think of themselves and Armed Forces personnel as part of one team, bridging the civil–military gulf. The charity targeted the sporting and pop worlds, gaining the backing of cultural icons such as David Beckham, Princes William and Harry, and national team members. From being unappreciated, Armed Forces personnel – particularly soldiers serving in Iraq and Afghanistan – became one of the most celebrated parts of Team GB.

From late 2007, the public not only gave moral and material support to the Armed Forces, but was eager to demonstrate their backing. Parry reports that the reaction by soldiers was 'huge':

> People were coming back from Afghanistan and Iraq and saying, 'Wow, something has changed. Everyone is now respecting us for what we do'.[119]

Following the example of other causes, Help for Heroes sold wristbands, an idea of the Parrys' daughter. The wristbands enabled supporters to show their support for the wounded – and by extension to the Armed Forces. Parry says:

> Richard Dannatt wore it immediately to Downing Street. Gordon Brown was seen wearing one; the First Sea Lord was wearing one. Anyone who wasn't wearing one would have to answer 'why?'[120]

Unlike the poppy, customarily only worn for a few weeks, the wristband signals the wearer's support at any time. Many regiments now have individual wristbands. However, with the debate about 'poppy fascism' started by broadcaster Jon Snow in 2006, what might be called the accessories of compassion can be symbolic of far more than a demonstration of moral and material support.[121] Liebenstein identifies the 'non-functional utilities' of consumer products: for Goffman, accessories such as poppies and wristbands are a means of conveying information, chosen by the individual – who can be performers – wanting to express an aspect of him- or herself. Although individuals as performers live in the moral world and are concerned with moral matters, they do not necessarily have moral concern,

118 Ibid.
119 Ibid.
120 Ibid.
121 BBC News, TV's Snow Rejects Poppy Fascism, 10 November 2006.

sometimes being more interested in the 'amoral issue of creating a convincing impression'.[122] The communicative function of the Help for Heroes wristband varies with the wearer: the head of the Army wearing it in Downing Street sends out a political message. With bracelets and bangles considered until recently a feminine accessory, the charity's Help for Heroes band allows a male wearer to associate himself with macho British military 'heroes'. The unseemly annual race by those in the public eye, particularly politicians, to wear a poppy as early as possible is less to do with the cause of Forces' welfare but more about being perceived to care – and by implication to care more than those who delay.

Homecoming parades have become a recent part of Britain's cultural landscape, reflecting the post-migration change in the public attitude towards the Armed Forces. Following a call by CGS Dannatt in September 2007 for local authorities to take the lead in organising them, both the *Daily Telegraph* and *The Sun* actively championed parades.[123] In October 2007, the Somme company of the Grenadier Guards returning to London from Afghanistan were reported as marching through empty streets: a similar lack of interest was recorded when the 3rd Logistics arrived home in Abingdon. Despite the MoD reportedly stating the issue was 'a matter for councils', Veterans Minister Stephen Twigg wrote to the Local Government Association urging them to 'see what can be done to show troops that we value the job they are doing for us'. The parade phenomenon quickly gained traction, culminating in 20,000 members of public turning out in Leicester to welcome home returning 9/12 Lancers: two days later 10,000 people reportedly watched a parade in Northampton.[124] Such visible public support indicates the closure of the civil–military gulf in terms of public appreciation.

The commemoration of the war dead is central to the Military Covenant. However, the rituals surrounding commemoration emerged after the First World War, a time when social equality in sacrifice was fostered by the government, in part through its ban on repatriation. Before then, Britain did not view the war dead – certainly not ordinary soldiers – as glorious: bones from Waterloo were ground for fertiliser. Few war memorials existed. Today, commemoration has become the subject of academic inquiry. King examines the politics of remembrance and the iconography of war memorials; Gregory explores the rituals surrounding the two-minute silence; Connolly focuses on commemorative practice in the East End, in particular the phenomenon of wayside shrines that originated in Hackney in 1916.[125] Whether around a village war memorial or at the Cenotaph, the invented traditions of Remembrance have been, as King observes, 'a source of increased communal solidarity'.[126] For Anderson the 'void' cenotaphs are 'saturated with ghostly

122 Goffman, p. 243; Liebenstein, 1950.

123 Adams and Simpson, 26 September 2007.

124 Peake, 17 March 2009.

125 Alex King, 1998; Gregory, 1994; Connelly, 2002

126 King, 1998, p. 60.

national imaginings'; the war dead helping to cement the imagined community that stems from shared nationalism.[127]

The continuity of the rituals and symbols of Remembrance that date from 1918 have helped to reinforce the sense that, as Fussell suggests, the Great War is an archetype, shaping notions of subsequent wars.[128] The claim by Alan Clark in 1961 that 'the First is as remote as the Crimean' might have been true for some of his generation, but half a century later the 'War to end all wars' with its familiar motifs – trenches, mud, gas, shellshock – carries on being fought in films, TV series and bestsellers.[129] This cultural focus enables the First World War to be embedded still further in the collective national memory – or the memory of memory. The end of 'living memory' of the Great War came in July 2009 with the death of its last surviving veteran, Harry Patch. The military and ceremonial aspects surrounding his funeral were somewhat ironic for someone who said war was 'organised murder' and Remembrance 'just show business'. During his last decade, Patch had become a living memorial, his life defined by his few months on the Western Front. Such was his moral status conferred by his link to the War and to the War dead, both the Prime Minister and the Sovereign acknowledged his passing.

The Military Covenant explicitly links the nation and today's soldiers with those who have made the ultimate sacrifice in other wars. The moral authority of such sacrifice is unassailable; consequently, association with it bestows moral authority by proxy. The reverence accorded to military sacrifice today is nothing new; Socrates articulated the Republic's need to honour its fallen:

> And then those who die bravely on active service we shall reckon men of gold …
> And treat their tombs with reverence and worship them as Guardian Spirits.[130]

As Gregory suggests in his exploration of the Silence, remembrance is dual-natured, allowing for public and private commemoration.[131] In 2009, the ninetieth anniversary of the Silence, a poll found 94 per cent agreed its observation was important, with the Legion stating it was 'as much about the Afghan generations as it is for the Armistice generation'.[132] The linkage of two groups of soldiers involved in conflicts almost a century apart reflects not only the potency of the Great War as an archetype, but also tacitly invokes its moral authority.

The Glorious Dead, inscribed on the Cenotaph, the focus of national sentiment, is an unambiguous message about the ultimate sacrifice – and, by inference, the cause for which that sacrifice was made. King observes that after 1918 'memorials

---

127 Anderson, 2006, p. 143.
128 Vernon Scannell, The Great War, Fussell, 2000, p. 321.
129 Clark, cited by Connelly, p. 2.
130 Plato (trans. Lee, 1987), p. 256.
131 Gregory, 1994, p. 18.
132 Ipsos MORI, Two Minute Silence, 2009.

and ceremonials took on as much a didactic as a commemorative character'.[133] The war dead have often been pressed into the service of the state, with memorials to their sacrifice emphasising their 'ethical triumph' rather than military victory.[134] Patch's death on 25 July 2009 coincided with the highest monthly casualty rate in Afghanistan. Inevitably, commentators made the link between Patch, the Great War and Helmand.[135] Following the death of eight soldiers in 24 hours on 10 July, an ICM poll for *The Guardian* and BBC's Newsnight showed support for British involvement in Afghanistan had risen – up 15 points from its level in 2006, despite 42 per cent being in favour of immediate withdrawal: however, by the end of the month a ComRes survey showed 52 per cent in favour of withdrawal.[136] While no causal link can be made be made with Patch and the drop in support for the war in Afghanistan, the findings suggest that the public was not prepared to be morally blackmailed by the sacrifice of soldiers in Helmand into supporting the mission. This aspect of the civil–military gulf remained unchanged.

The ritual surrounding repatriation that grew up after April 2007 in Wootton Bassett narrowed the civil–military gulf in terms of appreciation for soldiers and their sacrifice. The people of the town kept covenant with the fallen in an act of collective homage, reflecting how commemorative practice spontaneously evolves among a community. Similarly, the shrines to serving soldiers and the fallen that appeared in Hackney in 1916 gave impetus to the war memorials movement.[137] The evolution of the rite at Wootton Bassett – engendered by the introduction of mandatory repatriation – is another 'unignorable part of contemporary British culture' identified by Anthony King, along with the naming the dead in Parliament and posting obituaries on the MoD's website.[138] In contrast to the ritual surrounding the annual service of Remembrance at the Cenotaph led by the Sovereign in which political, religious and military leaders participate, Wootton Bassett could be described as the people's procession. As well as being secular, it was strikingly non-militaristic. King's observation that in recent commemorative practice, 'the domestic sphere has attained primacy over the state' was reinforced. While the coffins in Wootton Bassett were draped in the Union flag, State officials were absent. Senior politicians stayed away from the repatriations, as for the most part, did senior commanders. When uniformed Service personnel attended, they usually had a personal connection with the deceased. There were no military bands or military escorts for the cortege: the closest to any military ritual was when British Legion members lowered their flags. With the placing of flowers on the hearse, the dead were publically claimed by their family, extended family and friends. Wootton

133 Alex King, 1998, p. 173.

134 Ibid., p. 176.

135 E.g. Riddell, *Daily Telegraph*, 27 July 2009: Parsons, *Daily Mirror*, 1 August 2009.

136 Norton-Taylor et al., *The Guardian*, 13 July 2009; Morris and Sengupta, *The Independent*, 29 July 2009.

137 Connelly, pp. 25–35.

138 Anthony King, 2010, pp. 1–25.

Bassett also represented how the fallen are now 'personalized and domesticated', commemorated as family members 'fathers, husbands, wives, sons and daughters' rather than as soldiers.[139] At Wootton Bassett the human cost of conflict was apparent: death was not glorious but a loss to those who knew the dead soldier.

Wootton Bassett was tangible evidence of the Covenant between the military and the public. The High Street crowds could be seen as standing proxy for the nation, representing its appreciation of sacrifice. However, the town also represented how, in order to close the civil–military gulf by giving its moral support for soldiers and show its appreciation for sacrifice, the public had to separate the men and women from the missions. Local people shied away from politics, expressing reservations about the broadcast of BBC1's *Question Time* in December 2009 from the town. Mayor Steve Bucknell said:

> We've been careful throughout this process not to get involved in the politics of the war. These repatriations are simply about the soldiers and their families and the support we give to the Armed Forces.[140]

While King suggests that twenty-first-century acts of remembrance may encourage support for military campaigns by stealth, the majority of local people who attended the repatriations stated they were supporting the fallen, rather than the missions in which they fell.[141] One said: 'I am totally against the war – the lives it has cost and the meaningless efforts that have been put into it are unnecessary'. For two years, much of the television coverage was confined to regional bulletins, however in July 2009 the BBC national news led on the repatriation of eight Servicemen from Afghanistan.[142] The reporter said that Wootton Bassett was not about war aims or resources, but about 'the men who did not choose this conflict but were sent into harm's way to serve their country'. In contrast to the death of more than one million Servicemen between 1914–18, death in Iraq or Afghanistan is not described as glorious: the lack of autonomy of the fallen is stressed along with their separation from the war's aims. Although *pro patria mori* is reinforced by the use of the flag, there is no suggestion of *dulce or decorum est*. The public closed the civil–military gulf by acknowledging sacrifice, even if it did not agree with the cause for which sacrifice was made.

Conclusion

From 2003, the unpopularity of the Iraq intervention and the tepid support for the Afghanistan mission drove a wedge between soldiers and public. However, the

139 Ibid.
140 BBC News, Wootton Bassett Sheds its Tears for Soldiers, 10 November 2009.
141 Anthony King, 2010a, pp. 1–25.
142 Hewitt, BBC News, 14 July 2009.

civil–military gulf that concerned senior commanders and policy-makers, which came to be framed in the Military Covenant context, was never as wide as they perceived. Episodes of anti-military sentiment were exceptional and exaggerated, not least by the media and the National Recognition Report. Despite the public's regard for the Armed Forces being reflected by consistently favourable opinion polls, according to CGS Dannatt, soldiers themselves felt undervalued and unappreciated, considering the public indifferent to the challenges they faced. The fundamental principle of the Military Covenant – society's support in exchange for the offer of sacrifice – was going unfulfilled.

From late 2007, the Legion's *Honour the Covenant* campaign and Help for Heroes encouraged the public to make explicit its hitherto staunch but implicit support for Service personnel. Both charities were tangible expressions of the tenets of the Military Covenant and aided the concept's further migration and entrenchment in the civilian sphere. Support was further cemented from 2008 by the informal repatriation ceremony that grew up at Wootton Bassett, which became a globally renowned phenomenon.[143] The profound shift in public attitude and its corresponding fulfilment of the Military Covenant was inspired by the growing awareness of soldiers' sacrifice, but this could only be achieved by a conscious separation in the collective public mind of Service personnel from the military missions in which they were involved. Wootton Bassett's High Street crowds have been explicit about their rejection of 'politics', a reflection of the controversy that continued to surround the Iraq mission and the lack of enthusiasm for Afghanistan. Many interviewees suggest that the Armed Forces now enjoy unprecedented levels of support and point to displays of the public's affection, including homecoming parades and initiatives such as Tickets for Troops. If donations to Forces' charities are one metric of support for the military, after 2007 it was at an all-time high.

The Covenant's moral imperative of support in exchange for service and sacrifice provided a bridge between the military and civilian spheres that avoided the political controversies surrounding Iraq and Afghanistan. The civil–military gulf was narrowed to a gap between the two spheres, the civil–military norm in Britain. Through his deployment of the Covenant concept from October 2006, General Dannatt helped to generate the support the Forces' personnel came to enjoy. Media-friendly and ignoring constitutional proprieties, he appealed directly to the public, something about which he is 'unrepentant'.[144] The change in public attitude towards the Armed Forces from 2007 came about independently of government. Governments might try to promote cultural change through legislation such as criminalising 'hate', but active, positive support for any cause or group cannot be coerced. Describing it as 'out of kilter' in 2006, Dannatt believes the Military Covenant was being 'brought back into balance' two years later.[145] He says:

143 Reported in Japan and praised by President Obama.

144 Dannatt, interview.

145 Ibid.

'At least the people of this country were solidly behind us' – which implies that the government was not.[146]

146 Dannatt, 2010, p. 282.

Chapter 5
The Military Covenant and the Nation: Policy-Makers

Introduction

Immediately above the signatures of Prime Minister David Cameron and his Deputy Nick Clegg that concluded their foreword to the 2010 Strategic Defence and Security Review, was a commitment to the renewal of the Military Covenant; '… That vital contract between the Armed Forces, their families, our veterans and the country they sacrifice so much to keep safe'.[1] In January 2007, on board *HMS Albion*, another Prime Minister had stated that the covenant between the Armed Forces, Government and people had to be renewed: 'For our part, in Government, it will mean increased expenditure on equipment, personnel and the condition of our Armed Forces; not in the short run but for the long term'.[2] Tony Blair's reference to the Covenant is, with hindsight, significant. The first acknowledgement of the concept by a leading politician, it signalled that the process of migration from the military to the civilian sphere was underway and anticipated the future extension of the Covenant's scope to include the Armed Forces. More importantly, by giving it the Prime Ministerial *imprimatur*, Blair contributed to the Covenant's legitimisation. After all, he had called for it to be *renewed*, implying that it was a matter of historical fact.

The Prime Minister's acknowledgement of the Covenant was a tactical success for the Army. The Chief of the General Staff, Sir Richard Dannatt, had first publically invoked the concept in connection with soldiers' welfare and operational support in October 2006: 'The Army won't let the nation down, but I don't want the nation to let the Army down'.[3] Blair's speech indicated that the Government was reacting to a policy agenda set by the military, a reversal of a key civil–military relationship norm. From then until the 2010 General Election, the Labour government would be out-manoeuvred by the Army, which fought on the conceptual ground of its choosing: the Military Covenant.

In the context of 'Blair's wars' of choice as well as 'Brown's budgets',[4] this chapter explores the impact of the Military Covenant on policy and policy-makers, by assessing the material support offered to soldiers and the wider Forces'

1 SDSR, 2010, p. 6.
2 Blair, Portsmouth, 15 January 2007.
3 Sarah Sands, 12 October 2006.
4 To paraphrase Kampfner, 2004; Cornish and Dorman, 2009, pp. 247–261.

community before and after the Covenant's migration. However, soldiers, veterans and their families could well have had a sense of *déjà-vu*: most 'people issues' had been previously identified, if not tackled. The continuing debacle over defence procurement prompted Gray to observe: 'This seems to be the stuff of annual recrimination, "Why has the problem endured for so long?"'[5] Similarly, some of the welfare issues addressed in, for example, the 2008 Service Personnel Command Paper, had been considered by policy-makers almost a decade earlier. The Royal British Legion's *Honour the Covenant* campaign is examined, particularly in the context of its impact on policy-makers, at whom it was specifically aimed. This chapter will also consider the extent to which the Covenant became another means of party political advantage, undermining the previous broad consensus on defence issues.

Pre-Migration

Labour's perceived indifference to the welfare of Forces' personnel, symbolised by a broken Military Covenant, was an unhappy irony for the government. Notwithstanding later criticism, before the 2003 Iraq intervention it had made progress on people issues. 'The biggest defence review since 1957',[6] the extensive consideration of personnel matters in the 1998 Strategic Defence Review (SDR) was unprecedented. Although the *Policy for People* was criticised as 'the thinnest part of the SDR … largely a promise to give the matter attention',[7] it undertook to 'put our people first':

> Past defence reviews have concentrated on strategy and equipment, sometimes with insufficient consideration of people. This Review has given people their proper place at the centre of our plans.[8]

Among those people would be more women and ethnic minorities 'whose potential we have not fully tapped'.[9] Following the SDR, a central plank of Forces' personnel policy became, and remained, equal opportunities. The percentage of women in the Armed Forces went from 8.6 per cent in 1998 to 12.3 per cent in 2003, while that from ethnic minorities rose from 1.2 per cent in 1998 to 6.5 per cent in 2003.[10] The Defence Committee observed that equal opportunities policies had engendered a 'profound' cultural change within the Armed Forces.[11] However, as it suggested,

5 Gray, 2009, p. 15.
6 King's Forum, 1998.
7 Strachan, Memorandum, 15 June 2006 (a view shared by Dandeker).
8 SDR, 1998, paragraph 127.
9 Ibid.
10 MoD, 2003, Supporting Essay 5, paragraph 5.15.
11 HC 29-I, 2000–01, paragraph 35.

the case for ensuring that the Armed Forces better reflected civilian society through the promotion of equal opportunities was 'not just a moral or a political one, but an urgent practical necessity'.[12] Indeed, any shortfall in the provision of welfare to soldiers, their families and veterans was generally described in terms of the effect it had on recruitment and retention. For example, the issue of whether unmarried partners were entitled to Service Families Accommodation was 'not just a moral question; it is a real retention issue'.[13] During the Blair government's early years, policy-makers perceived the welfare of the Forces' community primarily as a practical issue or a statutory duty rather a matter of moral obligation.

Any fracturing of the Military Covenant begins with planning assumptions in successive defence White Papers. Even before the Iraq intervention, the Armed Forces were undermanned and defence underfunded, both themes recurring in Parliamentary defence debates. Along with a continuing commitment to Northern Ireland, the SDR anticipated that British Forces had the capability for either one major war-fighting deployment such as the Gulf War or two simultaneous peace support operations such as Bosnia. 'We would not however expect both deployments to involve war-fighting or to maintain them simultaneously for more than six months'.[14] In October 1997, the Army was undermanned by 4,500 trained personnel, with full strength not expected until 2004. By early 2001, the manpower deficit stood at 6,000 personnel, while the date to achieve the full manning target had slipped to 2008.[15] The Defence Committee warned that manning in the Armed Forces 'cannot be described as a crisis, but it is a chronic problem and one which shows little sign of responding to treatment'.[16] *The Strategic Defence Review: A New Chapter* of July 2002 acknowledged that the Britain's Service personnel had been 'working at, or near, or in some cases beyond the boundaries of what had been planned in the SDR for some considerable time now'.[17] The problem of under-manning and resulting overstretch was exacerbated once Operation Telic was underway.

Between the SDR and Iraq, a number of major policy initiatives concerning people issues were developed, linked to two themes that were to become central to policy after Covenant's migration: better cross-Governmental coordination and the principle of 'no disadvantage'. The Service Families Task Force was created in 1998 'to address the issues caused by the mobility of Service families that are outside the direct control of the MoD', including schools' admission policy. The Armed Forces' Overarching Personnel Strategy (2000, revised 2003) was 'based on the "cradle to the grave and beyond" philosophy'; that is before recruitment and beyond retirement.[18] Its family policy aimed

12 Ibid., paragraph 55.
13 Ibid., paragraph 151.
14 MoD, 1998, paragraph 89.
15 HC 29-I, 2000–01, paragraph 61.
16 HC 29–1, 2001, paragraph 12.
17 MoD, 2002, vol. 1, paragraph 61.
18 HC 29-II MoD Memorandum.

> ... To manage with care the dependent families of Service personnel and to take their needs into account in the formulation of Service personnel policy so that they are *not disadvantaged* [my italics] in comparison with their civilian counterparts and, subject to the operational requirements of each single Service, the demands on them are not unreasonable.[19]

Following their study into provision for veterans, in 2003 Dandeker et al. recommended better coordination between Government Departments.[20] With repeated recommendations for a more 'joined-up' approach to deliver people policies both nationally and locally, inertia among policy-makers regarding implementation can be inferred.

With the reframing of people issues in the context of the Military Covenant, came accusations that the well-being of the Forces' community was not a government priority. However, former Labour ministers reject this. Kevan Jones MP says: 'Nothing can have been further from the truth. We did a lot'.[21] He points to the Armed Forces Compensation Scheme, the Veterans Agency and the Operational Welfare package as some of Labour's early initiatives. He adds: 'We inherited an absolute mess. Army accommodation was crap'. Adam Ingram points to cuts in the Army training budget throughout the 1990s and to housing that was 'truly appalling', some in such a poor state that rent was not claimed from soldiers. 'On a trip to Gibraltar I was told not to stand too close to the side of a building as the windows were falling out'.[22] He maintains that the Conservatives had neglected welfare ever since Frontline First and there was a 'far better focus on personnel issues' under Labour, an observation with which some senior commanders have agreed.

The Military Covenant impelled policy-makers to overcome Whitehall and inter-Departmental inertia. Ministers across the political spectrum have expressed their frustration with the public sector and its ability to implement policy, most memorably Tony Blair who complained of 'scars on his back'. As Richards et al. note, in recent decades, the multiplicity of agencies and quangos, as well as the contracting-out of policy implementation to the private sector, has reduced ministerial control.[23] Taking over from John Spellar MP, who introduced the Service Families Task Force, former Armed Forces Minister Adam Ingram abandoned cross-Departmental Ministerial meetings because 'there was no real benefit accruing from them': instead, issues would be dealt with as they arose with the relevant Department. He says:

---

19 MoD AFOPS A-25.
20 Dandeker et al., 2003, p. 134.
21 Jones, interview.
22 Ingram, interview.
23 Richards et al., 2008, pp. 488–498.

> Whether it was the health care, dental care, education, special needs requirements, you then realise just how difficult it was to make Departments move; Ministerial diktat, if that's the right word to use, didn't quite apply. You couldn't just say, 'do it'. It didn't happen that way, partly because you had local authorities, or National Health boards and they were remote from the power of the Minister ... Wherever I knew lay-down was taking place, all agencies of Government, central and local, should have said, 'This is good news for our area, this is of economic benefit to our area, there's a Battalion coming in, good news, and here's all the things we're now going to do for them'. But it didn't quite work that way, and I found that increasingly frustrating.[24]

A former MoD official suggests that, although the rationale behind the greater integration of defence-related medicine with the existing NHS provision was 'absolutely logical', in practice there were difficulties because of Departmental structures, in part because of the delegated and diversified system for health compared with the central system for defence.[25] 'We'd have very good meetings [at senior inter-Departmental level] but that wasn't necessarily what was happening on the ground with this or that Health Authority'. He observes that other Departments and the Cabinet Office viewed the MoD as being 'too separate' and was told, 'You don't play a sufficiently integrated game with the rest of Whitehall'. Defence would have to implement Government policies for education, medicine and health and safety for which there was no extra funding.

> If there had been an education reform with a lot more money for teachers or class sizes, we were expected to apply those in the defence community but we were not given the extra resources to do so in our budgetary settlements. That was going to civil Departments. So I have to would look for £10m or £15m, which might not sound very much, to, for example, raise the certification level of heavy goods drivers in Germany.[26]

Former Defence Minister Kevan Jones MP emphasised the difficulties of trying to coordinate cross-Departmental delivery, citing 'hellish resistance across Whitehall' to implementing the 2008 Command Paper.[27]

Before the migration of the Military Covenant, policy-makers rarely placed the welfare of the Forces' community in the context of moral obligation. One exception, briefly, was Gulf War Syndrome. In July 1997 the Government stated that it had a 'debt of honour to those who have served their country in the Armed Forces'. The 2003 Defence White Paper, *Delivering Security in a Changing World*, gave little impression of any understanding of the unique nature of military

24 Ingram, interview.
25 Non-attributable, interview.
26 Ibid.
27 Jones, interview.

service. An infelicitous use of language implied a lack of empathy with the Armed Forces' ethos: 'As we continue to demand more of our people, we owe it to them to offer a fair employment deal'.[28] A 'fair employment deal' fails to convey any sense of moral obligation in exchange for the offer of, or actual, sacrifice; it could apply to any public sector worker. The Government's implicit message was that Armed Forces' personnel – many of whom were in Iraq – were not doing anything exceptional. In 2005, as a consequence of Deepcut, the MoD would be castigated by the Defence Select Committee for adhering too rigidly to the difference between its moral obligations and its legal requirements, which impeded a development of a duty of care policy:

> By maintaining a dividing line between its legal and moral obligations, MoD is open to criticism that it considers obligations that are not legally enforceable to be less important.[29]

Unwittingly, the Committee anticipated the apparent disregard of policy-makers for the moral obligations inherent in the legally unenforceable Military Covenant.

Migration

The first of 'Blair's Wars' – the interventions in Kosovo, Sierra Leone and even Afghanistan in 2001 – were fairly minor efforts compared with what came later. Indeed, even as deliberately low-key preparations were being made for a possible intervention in Iraq, the primary focus of military activity was the fire-fighters' dispute, which along with the fuel protests of 2000 and the 2001 foot and mouth outbreak, saw the Army aiding the civil authorities. Consequently, the offer of sacrifice by soldiers failed to resonate with the public as it would after 2003.

The Military Covenant reframed the debate about people issues, placing them in a moral context, and provided fresh impetus for policy-makers to address them. Although senior commanders point to the progress the Labour government had made after 1997, the military operations in Iraq and Afghanistan crystallised many of the challenges confronting policy-makers in connection with defence before 2003 and brought new ones. Former Army Families' Federation Chief Executive Julie McCarthy suggests that without Iraq and Afghanistan, the Military Covenant would never have gained the traction it did, a view shared by many interviewees.[30] In his 2007 speech, the Prime Minister stated that Government's material support had a moral aspect: the standard of force protection, accommodation or the air bridge, he suggested, 'symbolize the respect and gratitude' for the work then being undertaken by Armed Forces personnel. As he pointed out, 'ten years ago none of

28 MoD, 2003b, paragraph 5.35.

29 HC 63–1, 2005–06, p. 5 and paragraph 30.

30 McCarthy, interview.

this would have mattered so much', implying that combat operations gave these issues more saliency. However, by then, British Forces had been in Iraq for almost four years. It was only when people issues were placed in the Covenant context, first by General Dannatt and then by the Royal British Legion with its *Honour the Covenant* campaign, that policy-makers were compelled to address them with greater resolution.

While some progress was made after 1998, soldiers themselves pointed to the gap between the Review's rhetoric and the subsequent reality, exacerbated after the 2003 Iraq intervention. On becoming CGS, Dannatt considered that his 'primary task' was to focus upon them:

> … There was a litany of issues that demanded attention, in addition to those associated with the provision of equipment on operations – including among others, 'people issues', such as housing, pay and medical provision for our wounded.[31]

The people issues associated with the Covenant – which the CGS would describe as 'out of kilter' and 'out of balance'[32] – have varied since the concept began its migration in October 2006. Recently, the Continuity of Education Allowance has become salient, despite not being on the list of ten issues in the MoD's Military Covenant Factsheet of February 2008, which included compensation for the injured.[33] Although the welfare issues that became identified with the Covenant has varied, one constant has been the provision of healthcare, mental and physical, for the wounded, veterans and reservists.

With the British military effort in Iraq and Afghanistan relying on £18 billion pounds of ad hoc Urgent Operational Requirement payments and Treasury contingency funds, the MoD and the Treasury would have been reminded that 'the sinews of war are endless money'.[34] However, even before March 2003, the defence budget was under pressure. As the King's Forum suggested in 1998, although the Strategic Defence Review was 'genuinely' foreign policy-led, it was also 'inevitably resource-constrained', failing to address 'the problem of the mismatch between commitments and resources'.[35] As Ingram observes: 'The SDR was never funded'.[36] Former MoD Permanent Under Secretary, Sir Kevin Tebbit told the Chilcot Inquiry that the defence budget was 'always under stress', with a shortfall of about £500m, even before the complexities engendered by a switch in Departmental accounting procedures in 2004.[37] As Ingram suggests, unforeseen

31 Dannatt, 2010, p. 266.
32 Dannatt, speech, July 2007 and interview.
33 MoD, 2008c.
34 Cicero.
35 King's Forum, 1998, p. 23.
36 Ingram, interview.
37 Tebbit – Chilcot, 3 February 2010, p. 3.

costs could be incurred: during an exercise in Oman, the SA80 rifle was found to jam. 'We got onto it quickly and we fixed it very quickly, but it was £80m which someone had to find from nowhere'.[38] Chalmers argues that Labour gave defence 'a low budgetary priority'. If the defence budget had grown at the same rate as the rest of public spending, he estimates it would have reached £61bn in 2008/09. In addition, the real costs of supporting both frontline capabilities and Service personnel rose by two per cent per annum over the previous two decades.[39] Cornish and Dorman suggest that Treasury officials enjoyed 'unprecedentedly' close oversight over defence policy and planning during the Blair era.[40] However, Blair himself generally escaped any criticism over the defence budget: instead Brown was castigated for it. His lack of sympathy with the Armed Forces as Chancellor would rebound on him when Prime Minister. The attacks by former CDS Lord Guthrie and former CGS Dannatt were particularly damaging to his reputation.[41]

The Labour governments of Blair and Brown failed to quell the suspicion they wanted cut-price conflict. 'Just In Time' acquisition, the Urgent Operational Requirement (UOR) process and reliance on contingency funding gave the impression that the Armed Forces were operating hand-to-mouth and that policy-makers were constantly being overtaken by events and playing catch-up. Former Defence Secretary Geoff Hoon said the defence budget for 2002/03 during the build-up to Iraq was 'under pressure', although the actual war fighting was expected to be supplemented through UORs.[42] Ingram argues the policy was a success: 'as the National Audit Office reported, it did work, it did deliver and it gave value for money'.[43] However, as Dannatt observed, what was urgent can become routine.[44] Problems arose because the maintenance for any extra equipment acquired via the UOR process came out of the core defence budget, compounding the problem of long-term underfunding. The MoD was dependent on the Treasury's contingency funds: in 2005–06, £2bn of the extra £8bn the Government needed was requested by the MoD, £1.2bn of that for 'peace-keeping' in Iraq and Afghanistan.[45] At the Chilcot Inquiry, Tebbit insisted that at no stage did the Treasury deny or withhold funding for military operations in Iraq, his testimony reinforced by both Prime Ministers. Brown stated:

38 Ingram, interview.

39 Malcolm Chalmers, The Lean Years: Defence Consequences of the Financial Crisis, in Codner and Clarke (eds), 2011.

40 Cornish and Dorman, 2009, pp. 247–261.

41 E.g. Lord Guthrie, Gordon Brown Did Not Give All We Asked For, *Daily Telegraph*, 6 March 2010; Richard Dannatt, As Chancellor Gordon Brown Did Not Understand Defence, *Daily Telegraph*, 2 March 2010.

42 Hoon, Chilcot, 19 Jan 2010, p. 128.

43 Ingram, interview.

44 Dannatt, Chilcot, 28 July 2010, p. 26.

45 HC 980, 2006.

> When you are fighting a war the first priority has got to be to make sure your soldiers and Armed Forces are properly equipped ... Every request that the military commanders made to us for equipment was answered. No request was ever turned down.[46]

Blair was equally robust about the level of support given to Service personnel in Iraq:

> My view, very, very strongly, is, when you are asking your Armed Forces to go into these situations, you put everything to one side other than making sure that they have the equipment they need and they have the finance to back it up.[47]

However, nothing would dispel the impression among the public that the Labour Government's attitude to financing the Forces was one of 'muddling through'.

Following migration, the perception that Whitehall penny-pinching led to avoidable deaths and injuries of Armed Forces' personnel, came to be seen through the Military Covenant prism. Defence Secretary, Bob Ainsworth, summarised public opinion in early 2008:

> An underlying allegation has been made from numerous quarters that deep down the Government do not really care about the men and women in the British Armed Forces, that the Covenant between the Government and the Armed Forces is broken and that we are sending people into harm's way without due care and support.[48]

The Government was damaged by the public's perception of miserliness in connection with Force protection, first during the early stage of Telic with a controversy about body armour, and then in Herrick, where the apparent helicopter shortage became emblematic of policy-makers' attitude. Abstruse arguments such as the switch to resource-based accounting within the MoD in 2003, failed to capture the public imagination as much as the moral argument that those who are offering sacrifice should be properly supported. A National Audit Office report that examines the acquisition of eight Chinook Mark3s, fails to pack the same emotional punch as media reports that a soldier died because of lack of helicopter lift.[49] Justified or not, Labour policy-makers would be accused of a betrayal of British troops on a scale unprecedented since the First World War, with the Military Covenant used to put the message across.[50]

---

46 Brown, Chilcot, 5 March 2010, p. 107, p. 115.

47 Blair, Chilcot, 29 Jan 2010, p. 167.

48 Des Browne MP, HC deb, 10 Jan 2008, col 568.

49 Major Matthew Bacon, September 2005, cited by Page, 2007, p. 338, National Audit Office, HC512, 2007–08.

50 The Army's responsibilities regarding Force protection – Chapter 6.

Policy-makers failed to uphold the Military Covenant by their tepid moral support for the missions. Military doctrine put the onus on the government rather than the Army to act as a moral judge and ensure the legality of an operation. *Soldiering* is unequivocal:

> The chain of command, from the Government downwards, is responsible for articulating and sustaining the morality and justice of the cause in question, based on the enduring ethos of the British Army, and the particular circumstances of each operation. Only on this basis of absolute confidence in the justice and morality of the cause, can British soldiers be expected to be prepared to give their lives to others.[51]

While not specifically examining the tenets of *jus ad bello* and *jus in bellum*, *Soldiering* makes explicit that just cause is a prerequisite for soldiers risking their lives. The cause must be worthy of the sacrifice that he or she is called upon to make: this is 'the spiritual basis of morale'.[52] Sebastian Roberts observes:

> There is nothing worse for a soldier than to be on an operation whose legitimacy is questioned very deeply, and whose worth is questioned by the people who've sent us.[53]

Although Parliament approved the Iraq intervention, the absence of a second UN resolution reflected international doubts about the mission's legitimacy. Its legality was called into question with the wavering opinion of the Attorney General Lord Goldsmith and the resignation of former Foreign Secretary Robin Cook, who stated that the mission had neither 'international authority nor domestic support'.[54] In January 2003, CDS Admiral Sir Michael Boyce told the Prime Minister he would require 'an assurance of the legal basis of the conflict', a request that he repeated two months later on the eve of the intervention 'formally and explicitly'.[55] Roberts' observation about the legitimacy and worth of operations goes to the heart of why policy-makers associated with Telic and Herrick lost the confidence of many soldiers.

Reinforced by soldiers' service and sacrifice in Iraq and Afghanistan, the deployment of the Military Covenant by the Army further weakened the Labour government's already threadbare moral claims in connection with its foreign and defence policies. The Labour Government had consciously sought to break with the past in the international as well as domestic arena, embracing the concept of humanitarian interventionism. According to the King's Forum, the link between

51 ADP, 2000a, paragraph 0304. Army Doctrine Publications. Ministry of Defence. © Crown copyright 2014.

52 Ibid., paragraph 0302, section 4.

53 Roberts, interview.

54 Cook, 18 March 2003, p. 14.

55 Boyce Chilcot, 3 Dec 2009, p. 88 and witness statement, released 27 Jan 2010.

defence and intervention had been 'perhaps the most fundamental philosophical issue' within the Strategic Defence Review.[56] As Daddow and Gaskarth suggest, the government had 'a belief in the utility of armed force for moral ends'.[57] In 1999 in Chicago, during the Kosovo intervention that is regarded by McCormack as 'the highpoint of New Labour's ethical foreign policy',[58] Blair outlined the Lawrence Freedman-influenced Doctrine for the International Community that set out the case for intervention. As his memoirs later observed: 'It was an explicit rejection of the narrow view of the national interest ... it drew predictable criticism for making foreign policy a moral cause'.[59] In Chicago he stated:

> We cannot turn our backs on conflicts and the violation of human rights within other countries if we still want to be secure. ... Now our actions are guided by a more subtle blend of mutual self interest and moral purpose in defending the values we cherish. ... The spread of our values makes us safer.[60]

With the justification for the interventions in Iraq and Afghanistan shifting as the missions progressed, doubts were compounded about their legitimacy, legality and worth. Initially both were sold to the public on national security grounds: according to Blair, the issue of weapons of mass destruction was the *casus belli* in Iraq.[61] When they turned out to be non-existent, a humanitarian and values-based case was made instead. Similar incoherence characterised the mission in Afghanistan after 2006. Betz and Cormack argue that 'Britain has never had a clear political aim or an overall plan in Iraq or Afghanistan'. The 'strategic drift'[62] further undermined the credibility of the Government's moral case for war – and the wars' worth. A conflict based on 'our values' was impossible to reconcile with Guantanamo, Abu Ghraib, Breadbasket, rendition and torture. In 2009, weeks after the fraudulent Presidential election in Afghanistan, one MP denounced leading members of the Karzai regime as war criminals who should be in The Hague.[63] The 'sexing up' of the September 2002 intelligence dossier, the death of David Kelly and the Government's associated attack on the BBC, the arrest of an anti-war protestor for reading aloud the names of dead soldiers at the Cenotaph and the ejection of an 82-year-old heckler from the Labour Party conference in 2005 hardly suggested that British values included free speech, while those same values did not appear

56 King's Forum, 1998, p. 33.

57 Daddow and Gaskarth (eds), p. 5.

58 Tara McCormack, From 'Ethical Foreign Policy' to National Security Strategy: Exporting Domestic Incoherence, in Daddow and Gaskarth (eds), p. 108.

59 Blair, Chicago, 22 April 1999.

60 Blair, 2010, p. 248.

61 Ibid., p. 452.

62 Betz and Cormack, 2009.

63 Paul Flynn MP, 8 Dec 2009 HC 2WH.

to keep us safe on 7/7.[64] The contradictions between the moral rhetoric of policy-makers in connection with the interventions and the reality, both in the operational theatres and in Britain, further eroded the legitimacy of Telic and Herrick in the eyes of the public and its trust in the Government's grandiose moral claims.

The Governments of Blair and Brown appeared reluctant wholeheartedly to commit to a long-term military mission, first in Iraq and then in Afghanistan, conceptually, materially or morally. A Rumsfeld-like 'stuff happens' insouciance was adopted towards Phase IV stability in Iraq, the planning for which, suggested Freedman, a 'cavalier attitude' had been taken.[65] As Blair belatedly acknowledged:

> What is now known is that you have to go in as nation builders and you have got to go in with a configuration of the political and the civilian and the military that is right for a failed state situation.[66]

Citing Bosnia, Kosovo and Sierra Leone, former CGS Sir Mike Jackson said the Army had plenty of experience of being in a country with its infrastructure in ruins in the immediate aftermath of a conflict.[67] However, in the run-up to the Iraq invasion, the relationship between the MoD and DfiD appears as dysfunctional as that between the Defense and State Departments in Washington.[68] Joint Force Logistic Component Commander, Major General (retired) Tim Cross told Chilcot:

> I got no sense that across Whitehall there was any coherence in a pan-Whitehall perspective of what this was all about … I got no sense that diaries were being cleared as we prepared for these operations in Iraq.[69]

Similarly, overseeing Phase IV in Basra in 2003–04, Sir Hilary Synnott testified: 'It was quite clear to me that Whitehall as a whole was not mobilized'.[70] Former CDS Lord Boyce told Chilcot: 'I don't think the Treasury ever thought we were on a war footing'.[71] Similarly, Des Browne stated: 'Nobody advised or briefed me when I became Secretary of State for Defence, that what we were trying to do in Iraq was to win a war'.[72] The ad hoc Cabinet committee set up to deal with Iraq was ineffectual: Provisional Reconstruction Team leader Mark Etherington who arrived in Basra in April 2006 felt no one was leading in Whitehall. 'I read that Iraq was a "top strategic priority" for the Prime Minister, that there was "strong

64 BBC News, 7 Dec 2005.
65 Lawrence Freedman, Blair, Chilcot, 29 Jan 2010, p. 220.
66 Blair, ibid., p. 181.
67 Jackson, 2008, p. 390.
68 Outlined by Woodward, *Plan of Attack*, 2004.
69 Cross, Chilcot, 7 Dec 2009, pp. 31, 32.
70 Synnott, Chilcot, 9 Dec 2009, p. 50.
71 Boyce, Chilcot, 11 Jan 2010, p. 102.
72 Browne, Chilcot, 25 Jan 2009.

drive" to deliver. I saw no evidence of that at all'.[73] In addition to the lack of a clear strategic aim, Betz and Cormack suggest that neither mission 'had the coordinated government machinery that success requires'.[74]

In their dealings with a military that was involved in war-fighting at a pitch unparalleled since Korea, both Labour Prime Ministers appeared unengaged. Dannatt says:

> While I was CGS and Tony Blair was Prime Minister, we only had one conversation, about five weeks before he left. Given that we were fighting two wars and I was heading the organization doing most of the fighting, I'd have thought he might have had an interest.[75]

Blair's Chief of Staff Jonathan Powell and his foreign policy and defence adviser Nigel Sheinwald nodded through Dannatt's appointment as CGS. According to Powell: 'We agreed it was not worth consulting Tony about such a trivial subject'.[76] Although Dannatt is unsurprised,[77] it seems curious that when British forces were involved in two concurrent combat operations, the Prime Minister's advisers believed that the appointment of the new CGS was trifling. It reinforces the sense that all was far from well within the civil–military relationship, a situation that worsened after the CGS's interview with the *Daily Mail*. Powell states: 'We thought for a moment about sacking him but concluded that would just make him into a martyr'.[78] It seems far removed from the mutual respect and value codified by the Covenant. A similar attitude can be inferred from Gordon Brown's appointment of a part-time Defence Secretary, a post filled by five different incumbents in five years, which was reportedly regarded as near the bottom of Cabinet seniority.

The migration of the Military Covenant was engendered by policy-makers' failure to support Telic or Herrick materially and morally. Even in 2000, one MP questioned whether the government's grandiose moral ambitions were matched by its material commitment to the Armed Forces: 'Have we in fact underestimated the implications of having a more active foreign policy commitment to being "a force for good in the world?"'[79] With the military under pressure due to the protracted deployment in Iraq, leading to the Defence Committee in March 2006 noting the 'critical shortages in various trades in all three Services',[80] the Government's decision not to reassess its 2004 commitment to an expanded NATO operation in Afghanistan seems puzzling. The Defence Select Committee

73 Etherington, Chilcot, 9 July 2010, p. 8.
74 Betz and Cormack, 2009.
75 Dannatt, interview.
76 Powell, 2011, p. 89.
77 Dannatt, interview.
78 Powell, 2011, p. 270.
79 Dr Julian Lewis MP, 2000–01, HC 29-II Q27 6 Dec 2000.
80 Defence Select Committee, 2004–05, HC 822, paragraph 18.

was sceptical about claims concerning the possible impact of two operations on operational effectiveness:

> The MoD's confidence that the UK Armed Forces are not overstretched contrasts with what we heard from Service personnel on the ground [in Iraq] ... We believe that these concerns give rise to a fundamental question: are our Armed Forces structured, trained and equipped to fulfil the roles envisaged for them?[81]

According to its former Chief, the two simultaneous operations created the 'perfect storm' for the Army from mid-2006.[82]

While CGS Dannatt propelled the initial migration of the Military Covenant to the civilian sphere, the Royal British Legion's *Honour the Covenant* campaign entrenched the concept among policy-makers and opinion formers. Launched in September 2007, it focused on the welfare of Service personnel and their dependents in specific areas, such as compensation payments. Seeking changes to public policy, the campaign primarily targeted policy-makers and the media. The campaign was developed under the acting Director-Generalship of Frances Done, experienced in the workings of the public sector. The campaign was in its very early stages when it came across her desk in early 2007 and gradually coalesced during a series of weekly team meetings. The first task was to pin down a concept that was unfamiliar to many, including Done, who could however 'see immediately that this idea had the germ of something really important' and which seemed to tie in with a range of issues that were being tackled by the Legion, including delayed inquests.

> It struck me that the ideas around it were really important ... So bit by bit by bit we started to shape it ... Then there was a really good line, which started to come out – 'we can count on them, can they count on us?'[83]

The principle of reciprocity inherent in the Military Covenant was reinforced by the Legion's campaign.

> We perceived it as being a commitment between the public, whom the servicemen serve, but the only people who can deliver on behalf of the public, are the government. It's a call to action by the government, but it was [also] a call to action by the public, to say: 'We believe this is true, we believe there is this covenant, it's not a myth, we believe in it and we think people should be properly looked after, they should get help, so you've got to deliver it.'

81 Defence Select Committee, 2005–06, HC 1241, Summary.

82 Dannatt, Chilcot, 28 July 2010, p. 17.

83 Done, interview.

Robert Lee, the charity's Head of Media and Campaigns, says 'an inchoate sense of an obligation that needed to be met was very much alive'.[84] He suggests that the campaign was prompted by 'a raft of issues which made us feel that the nation, not just the Government but the nation, was not honouring service to a level of parity with the service that was being rendered'. Its working title, the Broken Covenant, was dropped by Done in favour of Honour the Covenant, which was 'a positive call to action rather than a criticism of the Government'.[85] Lee observes: 'We thought it was strained, we thought it was bruised, we thought in many cases it was not being honoured'.[86] There is an irony that the campaign which did so much to establish the Military Covenant in the civilian sphere kept its distance from the military by deliberately not using the term 'military'. There was also a question mark about 'covenant' and whether it should be used at all. According to Done:

> The whole concept of a covenant between the public and a Service person makes sense, but there was a lot of discussion about whether the public know what a covenant is, because, without being too superior, it's not a word in common use ... It did reflect the Military Covenant, but that wasn't the main issue because we knew that it wasn't widely known.[87]

The Covenant concept was so alien that the campaign's poster was assumed by many to be a promotion for the *Halo* X-Box game, whose chief antagonist is Covenant.[88] This led to 'a lot of social media chatter' according to Lee, who designed the poster:

> Any marketing theory will tell you, if you introduce a concept that does not resonate with the public, your campaign will fall flat. The genius in any successful campaign is putting words to something that people already feel and are already concerned about. That was certainly the case with the Covenant campaign.[89]

'Built into the architecture of the campaign' was lobbying by Legion members of their MPs, whether by postcards or at surgeries.[90] This grassroots action generated more than 22,000 interventions, of which about 10 per cent were meetings. Done's successor, Chris Simpkins took over shortly after the campaign's launch. He says:

> The feedback we got from our own membership on the *Honour the Covenant* campaign was almost, 'Thank goodness, about time we stood up and did

84 Lee, interview.
85 Done, interview.
86 Lee, interview.
87 Done, interview.
88 Confirmed by a straw poll of 20 Durham University undergraduates in June 2011.
89 Lee, interview.
90 Ibid.

> something like this, and we're up for this'. Which is why they all badgered their MPs.[91]

The Legion 'rediscovered its roots', having originally been set up as a campaigning organisation in 1921. 'We deliberately, very consciously changed our style. We were far more outspoken, far more upfront. We were prepared to be far more critical'.[92]

The self-proclaimed 'national custodian of Remembrance', the Legion could exploit its unimpeachable moral authority to pressurise policy-makers. As the *Guardian* observed, in offending the Legion, the Labour government could be judged as 'trampling on the memory of the 20th century's war dead'.[93] With a branch network throughout the country and a potential beneficiary community of nine million, policy-makers ignore the Legion at their peril. As Simpkins observes, few MPs will turn down a photo-call with their local Legion branch: Minister Des Browne told the Legion's Annual Conference in 2008, 'When the Legion speaks, governments listen'. In a pincer movement, a media campaign was conducted in parallel with the lobbying of Westminster. As the previous chapter has outlined, Help for Heroes and *Honour the Covenant* were launched within weeks of one another in 2007. Simpkins observes: 'It entered Parliament on a ball of fire. The response of MPs was swift'.[94]

The impact of Military Covenant's migration in the political sphere was to re-focus policy-makers' attention on the welfare of the Forces' community. Many issues emanating from the SDR's *Policy for People* were revisited. From September 2007, civilian advocates such as Legion added momentum to the process the military had started a year earlier. Before then, let alone during the years after the Covenant was first codified in 2000, Parliamentarians – certainly in the House of Commons – were not only unaware of the Covenant, but failed to suggest that any sort of particular obligation was owed to soldiers or to the wider Forces' community. An examination of Parliamentary debates, Defence Committee reports, written statements and oral answers between 2000–10 shows that until late 2007, the Covenant was an almost unknown concept. Lord Astor of Hever made the first reference to it in February 2007.[95] However, very few MPs showed much interest in defence. Even in June 2006, when British Forces were engaged on two fronts, only three members of the Government side were in the Chamber during part of a defence debate.[96] Until late 2007, debates themselves would follow a similar routine; the Government stressed ever-higher investment and improvements in equipment and welfare, while the Opposition painted a bleak picture of the problems faced by the Forces' community, with under-manning

91 Simpkins, interview.

92 Ibid.

93 Military Covenant: Poppy Politics, *The Guardian* Comment, 2 Nov 2007.

94 Simpkins, interview.

95 Astor of Hever, 19 February 2007, HL WA 171.

96 Andrew Robathan MP, 22 June 2006, HC Col 1545.

and the breach of harmony guidelines familiar refrains. In December 2006, the Defence Select Committee reflected upon the claim by its Chief that the Army was 'running hot'.

> General Dannatt spoke of the importance of the 'Military Covenant' between Service personnel and the Government and his concern that this would potentially be undermined over the longer-term if current levels of commitments were maintained.[97]

The concept's unfamiliarity is indicated by a footnote defining the Covenant: 'Service personnel making personal sacrifices while serving their country in return for being treated commensurately by the Government for this sacrifice'.[98] In June 2007, the first reference to the Military Covenant was made in a Commons' Defence debate: Nick Harvey cited an extract on morale from *Soldiering*, in the context of soldiers going absent without leave; a 'testament to the difficulties experienced by our troops'.[99] It could have been argued that these soldiers were breaking their covenant with the chain of command and their comrades.

The impact of the *Honour the Covenant* campaign, allied with the launch of Help for Heroes, was immediate at Westminster. This was reflected in the defence debate of 16 October 2007, throughout which Opposition MPs invoked the Covenant. 'Honouring' the Covenant varied: for one MP it involved homecoming parades and military-only hospital wards; for another it applied to the Forces' community.[100] The Covenant was also 'broken'.[101] The tone of the debate was, exceptionally, partisan and bad-tempered, but became a template for subsequent exchanges.[102] The Army's Covenant – the prism through which people issues began to be seen – became an Opposition weapon. Encouraged by their Whips, on 22 October three Conservative MPs, who had previously shown little interest in defence, asked the Armed Forces Minister 'if he would make a statement on the upholding of the Military Covenant'.[103] The Liberal Democrats' Opposition Half Day on 12th December was devoted to the Military Covenant, which it called upon the Government to 'renew'. A few weeks earlier in the House of Lords, the Covenant had been described as 'breaking down' and 'ragged'.[104] The attack on the Government by five former Service Chiefs during the debate indicated how politically divisive the issue of Forces' welfare was becoming.

---

97 HC57 (2006–07), paragraph 64.

98 Ibid., footnote to paragraph 64.

99 Nick Harvey MP, 26 June 07, HC Col 1560.

100 Julian Brazier MP and Bernard Jenkin MP, 16 Oct 07 HC Col 761.

101 Andrew Murrison MP, HC, 16 October 2007, HC Col 791.

102 E.g. the exchange between Bob Ainsworth MP and Andrew Murrison MP, 16 Oct 2007, HC Col 798.

103 Greg Hands MP, Ann Milton MP, Sir Michael Spicer MP, 22 Oct 2007.

104 Baroness Parks, HL col 933, Viscount Slim, HL col 992, 22 Nov 2007.

The Military Covenant broke the previous Party political consensus on defence. While the Iraq intervention had polarised the public, this had not been reflected inside Parliament, where it was a source of intra-Party disagreement rather than any conflict between Labour and the Conservatives. Iraq was considered to be a factor in the 2005 election, when Labour lost almost 100 seats, with votes going to the anti-war Liberal Democrats and, most notoriously, to the maverick George Galloway. However, Whiteley et al. suggest that Iraq and civil liberties were important for only about 3 per cent of the electorate.[105] Tony Blair observes that while a disproportionately large part of the media focused on Iraq as the major election issue, it 'played differently' for most voters.[106] Dorman questions the 'myth' of consensus in connection with defence in the 30 years after the Second World War, arguing that such an interpretation is a 'little generous'.[107] Despite this, none of the major strategic or defence decisions taken by successive Governments was reversed by their opponents. As Freedman observes, governments of both parties were anxious to reduce the proportion of GDP devoted to defence, down from 9.8 per cent in 1952 to 4.6 per cent in 1978.[108] Governments instead prioritised education and health, spending on which overtook defence in 1969/70 and 1972/73 respectively.[109] Capitanchik suggests that, the public has generally appeared 'indifferent' towards matters concerning the size, shape, equipment and deployment of the nation's forces, an attitude which was mirrored by Parliament.[110] The defence consensus had held since the mid-1980s, when the renewal of Britain's nuclear deterrent had been a general election issue in 1983 and 1987. Even if Labour were not the natural party of defence, the perception of the Government's indifference to the plight of Forces' personnel during a time of conflict probably began with Defence Secretary Hoon's skiing holiday on the eve of the Iraq intervention. This set the tone for what was to follow. Rejecting suggestions about his Party's indifference, Kevan Jones MP suggests: 'Most of the military come from constituencies like mine [Durham North]. It's not true Labour doesn't understand or care about defence'.[111] However, the public simply did not believe this.

From late 2007, the Military Covenant was not only civilianised but politicised. A narrative grew up in the media, encouraged by the Opposition, of the Government's 'betrayal' of soldiers and veterans. With the number of killed and wounded rising in Afghanistan – which to date numbers 448 fatalities, more

105 Whiteley et al., 2005, pp. 802–817.

106 Blair, 2010, p. 522.

107 Andrew Dorman, The Politics of Defence, in (eds) Croft et al., 2001, p. 104.

108 Freedman, 1999, p. 65.

109 Capitanchik, in Baylis (ed.), 1977, p. 280.

110 Ibid., p. 255.

111 Jones, interview.

than 100 in both 2009 and 2010[112] – the issue became more emotive. As soldiers' sacrifice became tangible, the nation's moral obligations to the military became more pressing, as codified in the Covenant. Wootton Bassett and Headley Court, symbolising the human cost of conflict, exerted huge moral pressure on policy-makers to ensure that Service personnel were indeed 'sustained and rewarded by commensurate terms and conditions of service'. In addition, the Legion's use of the imperative 'honour' implied that any inadequacies concerning people issues – which were to become Covenant issues – were seen by the public not merely in terms of poor management by policy-makers, but of dishonouring a quasi-sacred agreement.

The Covenant's influence on policy-makers was reflected by an arms race of concern about the Forces' community breaking out between all three main political parties in 2008. Understandably, after a decade of fairly feeble opposition, the Conservatives seized upon an issue which could be exploited and which put them on the right side of public opinion – and the press.. The Conservatives' *Report of the Military Covenant Commission* and *No Choice But Change*, by the Liberal Democrat MP Menzies Campbell reflect how, less than a year after the *Honour the Covenant* campaign, the Military Covenant had become embedded in policy-making. Written with input from RUSI, *No Choice But Change* extended the scope of the Covenant from welfare matters to broader defence policy. Sir Menzies explains:

> The most fundamental feature of the Covenant is that you don't ask people to do things that they can't do. You provide them with adequate equipment, you have a proper foreign policy, you have a proper assessment of military capability.[113]

The Conservative's Military Covenant Commission offered wide-ranging recommendations to 'repair the damage' to the Covenant. Commission chairman Frederick Forsyth, condemning the 'betrayal' of the Covenant, blamed Chancellor Gordon Brown for refusing soldiers necessary funding.[114] However, on the following page, it is noted: 'Any recommendations made by the Commission must be paid for within existing and planned defence budgets'[115] Andrew Murrison MP, who convened the Commission, says:

> It gave political momentum to the process. It was part of a range of interventions and initiatives that were taking place … All of us were encouraged by the efforts

112 *The Guardian* DataBlog, British Dead and Wounded in Afghanistan, Month by Month (ongoing).

113 Campbell, interview.

114 Frederick Forsyth in Conservative Party, 2008, Foreword.

115 Ibid.

> of others. The tragedy is that Labour was caught napping and dragging their feet on this.[116]

Both reports called for an Armed Forces Covenant involving all three Services. However, while Labour was condemned for the level of the existing defence budget, its opponents gave no undertaking to increase it.

The Military Covenant gave opponents and critics of the beleaguered Brown Government a weapon with which to attack it. Particularly after *Honour the Covenant*, criticism coalesced around the concept. Despite the many initiatives to improve welfare, after the Covenant's migration, Labour was always on the defensive. The Government did not help itself by having a part-time Defence Secretary in Des Browne, who, like Geoff Hoon, lacked any empathy with the military. In addition, the Government looked on the defensive with a so-called gagging order issued by the MoD to Service personnel to curb contact with the media: this was, according to the Chairman of the Armed Forces Federation, an attempt to 'stem the rising tide of criticism by members of the Armed Forces of the Government's failure to honour the Military Covenant'.[117] Similarly, proposed changes to inquests were perceived by the media as attempts to silence coroners who had condemned the inadequacy of Force protection measures. In 2008, the United Kingdom National Defence Association (UKNDA) was established, to campaign for 'sufficient, appropriate and fully funded Armed Forces'. Patrons included Lords Guthrie, West, Bramall and Inge. At a time of conflict, Ministers could not overtly criticise senior commanders – serving or retired – for their oversight of people issues.[118] Tensions between senior Government members and the Army added to a sense of Governmental dysfunction in defence. CGS Dannatt was perceived by many Labour supporters as a Tory sympathiser, with one observer stating that the antipathy between Browne and him was 'visceral'.[119] Former minister Adam Ingram notes the *Daily Mail*, *Daily Telegraph* and News International titles all supported the General.[120] Kevan Jones MP states: 'Blair should have sacked him when he made those statements about Iraq'.[121] When a defence minister believes the Head of the Army should have been sacked at a time when soldiers are involved in combat operations, all is far from well in the civil–military relationship.

The Brown Government failed to grasp the significance of the debate about the material resources being given to soldiers being recast in terms of the Military Covenant. Although Iraq and Afghanistan were Blair's wars, as Honeyman

116 Murrison, interview.

117 Richard Bartle, letter to *The Times*, 28 August 2007, p. 18.

118 Chapter 6.

119 Interview, unattributable.

120 Ingram, interview.

121 Jones, interview.

suggests, Brown could not distance himself from the consequences.[122] Perceived as a Chancellor who had determined domestic policy, any parsimony in connection with the defence budget would rebound on him: in the moral context of the Covenant, soldiers' sacrifice was blamed on a lack of financial support for the Armed Forces in general and Force protection in particular. With the Covenant's migration coinciding with, and conveying, soldiers' sacrifice in Afghanistan, Brown's timing as Prime Minister was unfortunate, especially as he had little empathy with the Forces. Seldon and Lodge observe: 'He thought the Military were always finding new hornet's nests and feared that the British would become a permanent army of occupation'.[123] A gifted communicator and unselfconscious about discussing values and faith, Blair had referred to the Covenant nine months ahead of the Legion's campaign to honour it. If Blair had still been premier, his record suggests he might have understood better the power of the Covenant's moral force and joined the debate on those terms. Lacking the showmanship of his predecessor, the less emotionally intelligent Brown said in another context that success is judged by 'practical actions for change'. As Dyson identifies, 'the financial crisis was to Brown what 9/11 was to Blair'.[124] Although Brown correctly insisted that money had been poured into defence, the Gurkhas' controversy was one example of many when the Brown government was on the wrong side of the argument involving soldiers' sacrifice.[125] In addition, the Covenant's migration changed the terms of the political debate in the Conservative's favour: honouring or breaking the Covenant superseded arguments about the illegitimacy, illegality and worth of the interventions in Iraq and Afghanistan, which most Tory MPs had supported. It is unsurprising that they seized the opportunity the Covenant gave them. As Patrick Mercer MP says: 'It was an easy political hit'.[126]

The Government produced two major policy documents in the wake of the Military Covenant's migration: *Report of Inquiry into National Recognition of Our Armed Forces* (the Davies Report) and *The Nation's Commitment: Cross Government Support to Our Armed Forces, their Families and Veterans* (the Service Personnel Command Paper).[127] Davies side-stepped matters such as poor quality housing that had become media and Opposition staples, instead focusing on identifying ways of encouraging public support for the Armed Forces. Despite the Report being signed off by Prime Minister Brown, Davies insists it is an independent document. Among many interviewees were the Duke of Edinburgh and the Prince of Wales, whose suggestions were 'extremely useful'.

122 Honeyman, 2009, pp. 85–100.

123 Seldon and Lodge, 2010, p. 297.

124 Stephen Benedict Dyson, New Labour, Leadership and Foreign-Policy Making after 1997, in Daddow and Gaskarth (eds), p. 81.

125 The Gurkhas' campaign – Chapter 7.

126 Mercer, interview.

127 MoD, 2008a, MoD, 2008b.

> If there were two people who had a record of doing what they could to support the Military and build bridges between the military world and civilian life it was them.[128]

Among the recommendations the Government accepted was an Armed Forces Day and greater engagement with the local public at military bases.

The Command Paper set out a cross-Departmental strategy to address the welfare of the Forces' community, involving not only Whitehall, but the devolved authorities and local councils. General Sir Timothy Granville-Chapman, then Vice Chief of the Defence Staff says:

> Ministers were very clear that they wanted to improve things in the personnel realm. A paper for Parliament was proposed. In the end it landed up being written by four of us round the table in my office. I was particularly keen on the notion that what mattered was that no Service person should be disadvantaged by virtue of serving. If we could establish that principle it could help those serving in relation to major issues (for example they should have a pay deal and TACOS that reflected the job they did and be properly equipped) as well as low level ones (not being disadvantaged by the fact that they did not have a postcode in advance when moving and so found it difficult to apply for schooling for their children or register with a doctor). All this went into Chapter One, together with the thought that, in some cases, it would be necessary for Service people to be given some preference in order that they were not disadvantaged.
>
> This was willingly accepted by the Minister (Bob Ainsworth) and the Secretary of State (Des Browne) and they, as shrewd negotiators (and with good work by their outer offices) pulled off the considerable feat of getting this accepted by the Devolved Administrations. The subsequent Service Personnel Command Paper (2008) was, in my view, the most significant step forward in personnel terms of recent times. I think the last administration's credentials in the people realm are good and it deserves more credit than it is customarily given. I think it is interesting that in all that has followed in the 'Covenant' work, the 'no disadvantage' rule has never been bettered and remains as the fundamental principle.[129]

The Command Paper provided a template for subsequent policy, including the recommendations in Strachan's Covenant Task Force report (2010) and Coalition policies associated with the Armed Forces Covenant (2011).[130] Work would be monitored by an External Reference Group (ERG), which would issue an annual report. The ERG, which included Forces' families' representatives, grew out of a small group used as a confidential sounding board once work on the Command

128 Davies, interview.

129 Granville-Chapman, note to author.

130 No disadvantage – Chapter 7.

Paper began in December 2007. Many interviewees praise Defence Secretary Bob Ainsworth, who they regard as unfairly pilloried, for pushing through the Command Paper and his subsequent commitment to implementing it. The Legion's Chris Simpkins says: 'Credit where credit's due. Bob and his successor, Kevan Jones, achieved actually more than I might reasonably have expected during the period of the last Labour Government'.[131]

Despite people issues being reframed in terms of the Military Covenant from late 2007, the Labour government rarely referred to it. Neither the Davies Report nor the Command Paper directly acknowledged the concept. There was no hint that the Covenant was torn, ripped, shredded or even a little creased: it did not exist. The omission is all the more surprising given the MoD's publication in February 2008 of the Military Covenant Factsheet which identified ten 'people issues' that were deemed to be within the Covenant's scope. The Command Paper accepted the Covenant's fundamental principle of support in exchange for sacrifice: 'Government has a moral obligation, on behalf of the Nation, to honour its responsibility towards its Armed Forces'; 'The relationship between Service people, government and the Nation is mutually supporting. Each must play its part in cementing that complementary relationship'.[132] While policy documents ignored the concept, academics including Edmunds, Forster, Tipping and McCartney focused upon it, raising awareness of it in Demos, RUSI and Chatham House, which published their papers. The Military Covenant became a 'portmanteau' term, its scope extended by the media and the Conservatives to cover everything from housing to the Nimrod crash.[133] Consequently, Labour's attitude seems as bizarre as Margaret Thatcher's insistence on talking about the community charge instead of the poll tax. Despite widespread agreement that progress was made in a number of policy areas, especially in connection with 'no disadvantage', after 2007 Labour always appeared to be reactive rather than proactive. Although some like Kevan Jones MP rightly contend they focused on practical policy rather than nebulous concepts, the Covenant came to symbolise soldiers' service and sacrifice – and, consequently, the Government's attitude towards that service and sacrifice. Through moral suasion alone, the Covenant changed public policy.

Conclusion

Following the 1998 Strategic Defence Review, the Labour Government cannot be accused of lacking policies regarding the welfare of the Forces' community. It made progress on the issue, particularly in relation to equal opportunities. Unsurprisingly, attention came to be focused on combat operations in Iraq and Afghanistan, both of which exceeded all planning assumptions and exacerbated

131 Simpkins, interview.

132 MoD, 2008a, pp. 2, 8.

133 Tipping, 2008, pp. 12–15.

existing problems regarding welfare issues and the underfunded defence budget. For almost a decade, the Government and Opposition had examined the same people issues, but did not view them through the Covenant prism until 2007. The concept gave fresh impetus to revisit them. The Covenant's impact can be measured by the number of policy initiatives it generated. However, improved cross-government co-ordination and the principle of 'no disadvantage' were both rehashed from almost a decade earlier.

After its migration to the civilian sphere, propelled first by General Dannatt and then, primarily, by the Royal British Legion, the Military Covenant became a shorthand term for the privations suffered by an overstretched military fighting wars on two fronts. While the Legion can be congratulated for its moral entrepreneurship in promoting the welfare of the Forces' community, after October 2007 the issue became politically partisan. As the Covenant's scope was extended into the area of Force protection, any breakage implied that the Government had soldiers' blood on its hands. Having taken Britain into a war whose legality and legitimacy many of the public questioned, the Labour government was out-manoeuvred and placed on the defensive by the Chief of the General Staff and a moral concept originating in military doctrine.

Chapter 6
The Military Covenant: The Army and the Individual Soldier

Introduction

With the migration of the Military Covenant to the civilian sphere from late 2006, the obligations that it had originally codified between the Army and the chain of command, to one another, and to the nation, were largely overlooked. Instead, the concept metamorphosed, coming to represent the bilateral relationship between the Government and the Forces' community; that is, the men and women from all three Services, their dependents and veterans. The three-way balance of mutual obligation inherent in the Covenant's original codification was lost. However, the civilian reinterpretation and promotion of the post-migration Covenant continued to serve the Army, providing it with a powerful means to protect and advance its institutional interests, while also engendering a better deal for soldiers.

This chapter seeks to establish the Covenant's impact on the Army as an institution and on its soldiers. It highlights how the Army rediscovered the concept and how it was used by senior commanders not only to muster civilian society's support for soldiers – if not for the combat operations in Iraq and Afghanistan in which they were involved – but also how it reframed 'people issues' concerning soldiers' welfare. Those issues – particularly concerning the healthcare of soldiers involved in frontline operations – became increasingly emotive among the civilian public and consequently evermore politically contentious.

The Military Covenant would not have become entrenched in the civilian sphere without the combat operations in Iraq and Afghanistan. This study suggests that the post-migration Covenant deflected civilian society's attention away from the totality of those operations to one aspect of them; namely, their effect on the welfare of individual Service personnel, with whom the Government was accused of 'breaking the Covenant'. From 2006, the increasingly politically partisan clamour in Parliament and the media concerning the state of Covenant – which the Army did little to discourage – drowned out questions about the rationale behind, and conduct of, the ongoing Operations Telic and Herrick.

This chapter examines whether the focus on the Covenant proved to be advantageous to the Army in the short term, postponing inquiries about the motives behind the institutional 'push' for expanding the military campaign in Afghanistan in early 2006 when the mission was far from accomplished in Iraq. The question about whether the Army in fact betrayed its Covenant with the nation by encouraging an intervention in Afghanistan out of institutional self-interest rather than supporting the national interest is explored. This chapter examines the decision-making process that allowed British forces to play a leading role in an expanded NATO mission in Afghanistan, which has been under increasing scrutiny.[1] As Operation Herrick turned increasingly sour, blame was attributed to a Labour government that had starved the Army of funds: Ledwidge suggests this is a variation of the stab-in-the-back myth, fostered in Germany after its 1918 defeat.[2] The Covenant's civilian champions, including Opposition politicians and the media, extended the Covenant's scope to encompass operational matters, including helicopter provision. However, claims about a 'broken' Covenant in this context raise a number of questions, including responsibility for defence procurement.

The Military Covenant licensed an unprecedented incursion into politics by senior commanders. With his deployment of the Covenant, General Dannatt entered the political arena. Although any Chief of the General Staff can be expected to defend institutional interests, especially at a time when soldiers are involved in two concurrent combat operations, this chapter examines how far Dannatt challenged existing civil–military norms. In doing so, it considers whether the Covenant has not just added a new dimension to analysis of the civil–military relationship in Britain but has altered that relationship in the long term.

The Covenant, the Army and the Individual Soldier

Soldiering: The Military Covenant was expected to have a small military audience. According to its author, *Soldiering* 'was only going to be a slightly graduate, slightly geeky thing' to be read by Major Atkins, rather than Thomas or Cadet Atkins.[3] Soldiers themselves, including many interviewed for this study, seemed either unaware of, or indifferent to, the pre-migration Covenant. None interviewed said it was a major topic in either the Mess or the NAAFI. As one observed: 'It's a good idea if anyone knew it existed'.[4] A concept that originated in the military only became familiar to many soldiers once it was promoted to, and by, civilian society.

Within the military sphere, *Soldiering* was always a background document to *Values and Standards of the British Army*, for which it provided the 'philosophical underpinning'.[5] The institutional primacy given to *Values and Standards* was reflected in the 2008 Aitken Report, which did not refer to *Soldiering* or the Military Covenant. Aitken stated that soldiers need 'not just to be able to recite the Six Core Values parrot-fashion, but rather to inculcate the practical adherence to them in all aspects of our lives, at home or on operations'.[6] In the context of the nation's

1 E.g. Clarke, RUSI, 2011, Chilcot, Haynes et al., 2011.
2 Ledwidge, 2011.
3 Roberts, interview.
4 Serving Officer.
5 Roberts, interview.
6 Aitken, 2008, p. 25.

obligations to soldiers serving in Iraq and Afghanistan, General Dannatt stated that the Army was not going to let the nation down, 'but I don't want the nation to let the Army down'.[7] However, when some soldiers let the Army and nation down by abusing Iraqi detainees, which some in the civilian sphere perceived as the biggest institutional moral failure since Deepcut, the Army framed their misconduct in the context of *Values and Standards*, while soldiers' reciprocal moral obligations to the Army and nation inherent in the Military Covenant went unmentioned. The continuing institutional importance of *Values and Standards* can also be inferred from the Army's public pronouncements following the Gage Inquiry report into the death of Baha Mousa. In stating there could be 'no excuse for the loss of discipline and lack of moral courage that occurred', CGS General Sir Peter Wall alluded to two of Army's six core values.[8] However, while institutional faith has continued to be kept with the values, they are open to subjective interpretation: one soldier's failure of moral courage is another's grassing up your mates. Perceived by military doctrine as a failure of moral courage, the 'closure of ranks' characterising the abuse cases is surely an inherent aspect of unit cohesion.

The legal component frequently calls into question the strength of the Army's Covenant with individual soldiers. In 2005, a former CDS stated the Armed Forces were under 'legal siege'.[9] The development of the Moral Component in the latter half of the 1990s coincided with the incursion of civilian law into the Army's professional space. This process examined by Rubin and Forster, who explore the juridification of the Armed Forces and the processes by which their legal autonomy was chipped away, not least with Human Rights law, which would add another dimension to the missions in Iraq and Afghanistan. Concurrently, the military justice system underwent radical change, culminating in the 2006 Armed Forces Act.[10] Legal action brought on behalf of both British military and Iraqi civilian claimants, including Private Jason Smith and Baha Mousa, has tested the jurisdiction of human rights law in connection with military operations overseas, as well as the concepts of combat immunity and Crown immunity. While the soldiers' edition of the Army's revised *Values and Standards* (2008) did not refer to the Covenant, the commanders' edition reflects the institutional understanding of the Covenant:

> The Army's loyalty to the individual is expressed in the Military Covenant. It manifests itself in justice, fair rewards and life long support to all soldiers.[11]

However, as a result of alleged or proven instances of soldiers' misconduct in Iraq, the Army found itself in the metaphorical dock, accused either of hanging soldiers

7 Sarah Sands, 12 October 2006.
8 MoD, 2011c.
9 Admiral Lord Boyce, HL 17 July 2005, c1236.
10 Rubin, 2002, Forster, 2012b, pp. 283–300.
11 ADP, 2008.

out to dry by needlessly prosecuting them, or of closing ranks and not doing so.[12] The emotional temperature was raised if the alleged incident was described as a 'war crime'.[13] Whenever the Army gets involved with the legal component, it seems it cannot win.

The Military Covenant's utility to the Army followed its migration to the civilian sphere. The relative unimportance of *Soldiering* appears, with hindsight, paradoxical, not least because it confronts the moral responsibilities of being a soldier; one who not only makes an offer of sacrifice but who can take life, something of profound relevance to many soldiers from 2003 onwards. As Roberts says, *Soldiering* concerns 'the fundamental significance of the individual soldier being a weapons bearer and having to make the right decisions about lethal force'.[14] However, neither the military nor the civilian spheres placed too much emphasis on soldiers' reciprocal obligations to the nation or to the Army, which the Covenant codifies. CGS Wall spoke of the reciprocity of trust between soldiers and nation, but did not refer to the Covenant in this context.

The post-migration concept of the fractured Covenant conveys the notion of soldiers betrayed by the nation for which they are prepared to sacrifice their lives. Civilian society's current support for the Armed Forces was generated not by strategic success, nor by episodes of military glory, but by the sacrifice of a comparatively small number of personnel. Soldiers also made the ultimate sacrifice in other recent conflicts, notably the Falklands. The number of military fatalities in Afghanistan in 2009 was 108, compared with 102 military deaths in Northern Ireland in 1972.[15] On 27 August 1979, 18 soldiers were killed at Narrow Water in Warrenpoint. Lagan Valley MP Jeffrey Donaldson suggests that, while Northern Ireland veterans do not begrudge the attention given to today's soldiers, 'There are a lot of people who feel a little forgotten'.[16] However, with repatriation and the accompanying media coverage, soldiers' sacrifice has been made highly visible to the civilian public. One observer suggests the rituals that have evolved around repatriation have lent nobility to military sacrifice and, by extension, the cause for which sacrifice is made. He suggests:

> The Generals quite like an acceptable level of casualties. A steady one a week or one every two weeks with a coming home ceremony. It shows it's a serious war, there is real fighting.[17]

12 E.g. 7 members 3rd Battalion Parachute Regiment in 2005, and Irish Guardsmen Joseph McCleary and Martin McGyng (Harding, *Daily Telegraph*, 9 June 2006); Shiner, *The Guardian*, 23 April 2007.

13 E.g. Col Tim Collins OBE.

14 Roberts, interview.

15 Box, 2009.

16 Donaldson, interview.

17 Interview – former government official.

The sacrifice of the few has led the veneration of the many, with all Armed Forces personnel enjoying unprecedented public esteem. Military sacrifice reinforced the legitimacy of the Covenant as a concept, not least because it reminded the nation of its obligations to soldiers who had made that sacrifice.

The civilian interpretation of the Covenant as fractured could undermine the military in the long term. Although an effective catalyst for marshalling support for the Armed Forces and renewing the civil–military relationship, one unintended consequence of the Covenant's migration was that serving and former soldiers, as a result of their military service, frequently came to be perceived as victims by the public. The trope of the broken Covenant promoted by Opposition politicians and some of the media conveyed a message of 'brave heroes betrayed'. The notion of soldiers' vulnerability is at odds with the war-fighting ethos of the British Army. Soldier-as-victim challenges military doctrine, which emphasises the primacy of combat and the fact – unpalatable to many in civilian society – that many young men want to fight, and enjoy it when they do.[18] Soldiers were eager to go to Iraq, Afghanistan or both. As Holmes suggests, it reflected the desire of professionals to see just how good they were at their job and to test their skills in the toughest environments. 'In part, the appeal of danger is enshrined in the male psyche, and many young men see combat as the ultimate challenge'.[19] As McCartney observes, Armed Forces men and women have been designated heroes or victims by the media.[20] Both terms are an inaccurate description of the majority of individual Service personnel, whether serving or 'veteran'.

Veteran-as-victim could be particularly damaging to the Armed Forces in the future because of its effects on recruitment and on serving soldiers who risk being stigmatised by the civilian public. Myths about veterans – concerning the disproportionate numbers who are homeless, in prison, vulnerable to mental illness, or who have committed suicide – have gained increasing currency. One recurring myth concerns Falklands veterans: the media and politicians regularly claim that more personnel have committed suicide than were killed in action.[21] Similarly, *The People* stated that 10,000 veterans were homeless, accounting for one-in-five rough sleepers based on claims made by the Soldiers Off the Streets charity.[22] In 2008, the National Association of Probation Officers suggested that veterans accounted for 9.1 per cent or 7,350 inmates in the prison population, and that 'most of the soldiers who had served in either the Gulf or Afghanistan were suffering from post traumatic stress'.[23] Not only is PTSD becoming a blanket term covering any adverse mental health condition, but with veterans taking an average of 13 years to 'present', the implication that they have been suffering

18 Highlighted in Sky TV's documentary series *Ross Kemp in Afghanistan*.

19 Holmes, 2007, p. 315.

20 McCartney, 2011, pp. 43–54.

21 McNab, 22 November 2008.

22 Harpin, 2011.

23 National Association of Probation Officers, 2008, pp. 1–2.

in silence underlines their victimhood. Looking for veterans to interview, the media often approach Veterans Aid CEO Hugh Milroy for candidates who fit their preconceived notions of hero-to-victim. Milroy's response to late-onset PTSD can be 'cherchez la vodka': he observes that some veterans put a now socially acceptable spin on their alcohol-induced midlife crisis.[24] Like many who have served in the Armed Forces, Milroy has a far less sentimental and reverential view of soldiers than many in the civilian sphere, who, if presented with evidence about soldiers' misconduct, tend not to challenge platitudes about 'bad apples' and a 'tiny minority'. Rather than focusing on the positive impact of military service for the overwhelming majority of the 20,000 personnel who leave the Armed Forces each year, some within civilian society – sometimes out of self-interest – have focused on its negative effects. Using unreliable statistics, they have contributed to the hero-to-victim myths, while a gullible media have colluded in myth-making, frequently taking a 'hero's' hard-luck story at face value.

In connection with the welfare of serving soldiers or veterans, the Army did little to discourage the view that the responsibility for any breakage to the Covenant lay with the Government rather than with the institution, the chain of command or indeed with soldiers or veterans themselves. In the decade under review, the Army did not challenge the current inclusive definition of a veteran – that is, someone who has completed one day's training. However, according to Dandeker et al., those recruits who either have served for very short periods or who have not completed training, are most likely to suffer social exclusion in terms of mental health, homelessness or unemployment.[25] While the adverse effect of military service has been exaggerated, the Army is not entirely exonerated from an institutional dereliction properly to support all soldiers' transition back to civilian life. Any such failure is a breakage of the Covenantal commitment, as laid down in military doctrine, to 'life long support to all soldiers'.

The Military Covenant's migration was, in part, prompted by a plea by the Army's senior commanders for the public's greater understanding and support for soldiers. Historically, one bridge between the military and civilian spheres has been the Reserve Forces. However, Reservists were not included in the original codification of the Covenant or in the 2005 revised version of the concept, despite their contribution to the Iraq campaign, where they numbered 20 per cent of the Forces. Since 2000, 29,000 TA personnel have been mobilised,[26] filling a skills and knowledge gap – for example in the IT sector – among Regulars. Former Deputy Commander Army Reserves, Major General the Duke of Westminster, says:

> We saved the Southern Iraq oilfields because we had a Sergeant who worked on a North Sea oil field and he knew which gaps to plug in the pipeline. There was a Royal Auxiliary Air Force Squadron Leader who worked for Shell. The

24 Milroy, interview.

25 Dandeker et al., 2006.

26 MoD, 2011d, p. 11.

> Americans were about to target the Oil Ministry in the Centre of Baghdad and he said: 'For God's sake, don't blow that up, they hold all the geological records going back 100 years for the whole of Iraq. I know because I've seen them'.[27]

Reservists' mobilisation since 2003 has revealed flaws regarding 'people issues'; initially, the most obvious was in connection with pay, with a gap between a Reservist's civilian and military salaries. 'We were mobilizing soldiers from the private sector and they were losing money serving their country'.[28] Doing battle at the Treasury on the Reservists' behalf, Westminster was told by an official that they 'don't do morality'. Reservists are offered the Proposition, 'the same sort of glue as the Military Covenant'.[29] This states:

> Defence will offer the challenge and reward which attracts people to volunteer, and undertakes to train and support them throughout their Service, including when mobilised and recuperating.[30]

However, the Reservists' exclusion from the Covenant's original codification highlights their longstanding difference from Regulars who, they believe, could be more appreciative of their contribution. According to Westminster:

> I tell Territorials you're not in the job to be loved, just kick on and get on with it ... There is a certain section of the regular Army who are quite astonished by what we do. They can't believe their job can be done by a part-timer and that's a difficult thing to overcome: that relationship needs to be handled really very carefully.[31]

Both the SDSR and Force 2020 indicate a greater future reliance on Reserve Forces with a shift in the Reserve-Regular balance. However, the 2011 Future Reserves report concluded that Britain's Reserves were 'in decline', in part because the Regular Forces have raided their funding:

> This has significantly eroded the trust that must exist between the Regular and Reserve components of our Armed Forces and is one of the critical issues which must be addressed.[32]

With Reservists expected to make up about one-third of the Army by 2020, steps must be taken by Regulars and the civilian sector better to 'value and respect' them.

---

27 Westminster, interview.
28 Ibid.
29 Ibid.
30 MoD, 2011d.
31 Westminster, interview.
32 MoD, 2011d, p. 13.

Between 2000–06, *Soldiering* and the Covenant concept were a comparatively overlooked aspect of the Moral Component, of little practical utility to the Army. This situation was to change dramatically with the Covenant's migration, initiated by General Dannatt in late 2006. Both then and subsequently he used the Covenant, with its moral principle of support in exchange for soldiers' service and offer of sacrifice, to communicate the pressures on the Army which was only 'just' coping, having reached a manning cliff-edge.[33]

> We were out of balance. When we're out of balance is when the engine scrapes and sparks fly and we have difficulty. We weren't organized for conducting two major operations.[34]

Although he would later invoke the Covenant in connection with the public's moral support for soldiers, he first linked it with material support: 'I was questioning the amount of resource being put into housing, welfare, pay, allowances, all those kind of things'.[35] However, in addition to commenting on 'people issues', the General criticised the government and voiced doubts about the Iraq mission: having 'kicked the door in … our presence exacerbates the security problems'.[36] What went largely unremarked were the devoutly Christian Dannatt's views on contemporary Britain, whose values were 'under threat':

> We can't wish the Islamist challenge to our society away … the Army both in Iraq and Afghanistan and probably wherever we go next, is fighting the foreign dimension of the challenge to our way of life.

Such on-the-record candour is unprecedented, disregarding the convention that only retired senior commanders make their views public. The Military Covenant underscores soldiers' service to the nation, to which they subordinate themselves. The Covenant invokes the moral values of responsibility and loyalty to the nation to which all soldiers are expected to adhere, as well as soldiers' service to the nation, to which, according to civil–military relations norms, they subordinate themselves. It is paradoxical that while initiating the concept's migration, the head of the Army challenged a prime civil–military relations norm, notably that of civilian political control.

General Dannatt would have been sacked if he had been a Chief in the United States, where there is far more rigorous policing of civil–military boundaries by analysts, the media and the Commander-in-Chief, reflected by President Obama's sacking of Stanley McChrystal.[37] Although Dannatt said he was speaking up for

33 Dannatt, interview and Norton-Taylor, *The Guardian*, Sept 2006.

34 Dannatt, interview.

35 Ibid.

36 Sarah Sands, *Daily Mail*, 12 October 2006.

37 Including Kohn, 2009; Myers and Kohn, 2007; Coletta, 2007.

the Army, his greater constitutional loyalty was to the Crown-in-Parliament. As Fisher observes, the Iraq intervention was democratically-elected government policy and approved by Parliament.[38] The attack on the Government by five former Chiefs in the House of Lords in November 2007, while acceptable according to British constitutional convention, did little to improve civil–military relations. In the US context, Kohn argues that 'the military's prestige and reputation for disinterested patriotism means that any public dissent by senior commanders weakens civilian leadership in the eyes of the public.[39] The 2006 'Revolt of the Generals' against Secretary Rumsfeld provoked reflection among civil–military relations analysts about former senior commanders' responsibilities.[40] Many agreed with Snider that the retired Generals 'had violated the tenets of their profession'.[41] In Britain, the framing of people issues in the moral context of the Military Covenant legitimised overt intervention in politics by senior commanders, whether serving or retired.

The period of General Dannatt's tenure as Chief of the General Staff was marked by unprecedented public friction in the civil–military relationship, best represented by the Army's deployment of the Military Covenant in the civilian sphere. Commanding an Army under pressure involved in combat operations on two fronts, neither of which enjoyed much public support, it is not unexpected that the CGS should fight its corner with ministers. By convention, this is done in private. However, such an approach could have well have been fruitless. According to one military observer, Dannatt's predecessor, Sir Mike Jackson went from being soldiers' 'good old Jacko' to 'you let us down, you bastard'.

> Troops were beginning to prod him in the chest and say, 'Why aren't you standing up for us?' That troubled him deeply because within the MoD he was standing up for them. He was being really quite robust in his discussions with civil servants and ministers. But he was obeying a code, which said you don't speak out in public. I'm sure in his handover to Richard Dannatt he said, 'Well, it's your call my friend but it seems to me that the path I've trodden and that my predecessor has trodden, has got us nowhere'.[42]

Many soldiers admire General Dannatt's loyalty to them, and to the Army as an institution. Major-General (retired) Peter Currie observes young regulars and officers applauded the CGS: 'Here's someone who cares about us'.[43] Officer D agreed:

38 Fisher, 2011, p. 254.
39 Kohn, 2009.
40 Donnelly (Moderator), 2006, *Rummy and His Generals*.
41 US Army War College, 2009.
42 Sharpe, interview.
43 Currie, interview.

> Although he was seen as being political and anti-Labour, he wanted to look after his people. Internally there was never any criticism of General Dannatt, he was acting in their best interests.[44]

After his *Mail* interview, Dannatt was more circumspect, speaking, for example, of the Covenant being 'out of balance'. The working title of the Legion's *Honour the Covenant* campaign had been the 'Broken Covenant', which was changed before the launch. Dannatt says:

> Where I disagreed violently with the decision of the staff of that campaign is that they talked about a broken Covenant. I said it's not a broken Covenant, it's out of balance. You won't get any support from me, in fact, you'll get criticism from me if you talk about a broken Covenant. Broken is something that's done, it's bad; I've failed in my job. Out of balance is a work in progress and I'm working to get this right. Honour the Covenant I'm comfortable with.[45]

As CGS campaigning for better terms and conditions for soldiers, Dannatt came to embody and uphold the Army's Covenant with the chain of command.

After the Covenant's migration, which put the welfare of the Forces' community on the political agenda, overt flouting of civil–military norms by senior commanders became routine. CGS Dannatt ignored constitutional conventions, to the disapproval of many interviewees for this study. One media commentator stated he had 'opened Pandora's Box:[46] Forster said the Covenant genie was let out of the bottle.[47] According to Finer's classification of military intervention in politics, the Army crossed the line from 'influence' into 'blackmail',[48] specifically moral blackmail. The Military Covenant became the means by which serving soldiers could exert moral pressure on a government and publically pass judgment on it. In the context of the Army's deployment of the Covenant, according to Feaver's principal-agent analysis the Army agent was clearly 'shirking' – not doing something to the principal's (i.e. the Government's) satisfaction,[49] with Dannatt in particular continuing to test the civil–military relationship throughout his tenure. In 2007, the CGS's Briefing Team Report was leaked to the *Sunday Telegraph*: soldiers were described as feeling 'devalued, angry and suffering from Iraq fatigue' and the Military Covenant was 'out of kilter'.[50] In 2009, Dannatt appeared to send a message about helicopter provision when he was flown in an American Blackhawk in Afghanistan, which did nothing to improve his relationship with

44 Officer D, interview.

45 Dannatt, interview.

46 Jenkins, *Sunday Times*, 15 October 2006, p. 16.

47 Forster, 2012a, pp. 273–290.

48 Finer (ed. Stanley), 2002.

49 Feaver, 2005, p. 60.

50 Evans, 19 November 2007, p. 14.

Ministers, who countered with publicising the 'pop-star' helicopter trips taken by senior commanders both at home and in Afghanistan.

The Covenant allowed senior commanders overtly to enter the political arena. Although the civil–military 'gap' is a centuries-old norm, as Strachan suggests, historically the Army has rarely been apolitical.[51] Andrew Robathan MP, suggests that as CGS, General Sir Mike Jackson politicised his role. 'He gave interviews defending [Labour] Government policy in a way which, in my opinion, was not appropriate for a senior officer'.[52] However, with the rapid adoption of the post-migration Covenant by the Opposition parties, the Army became enmeshed in Party politics, leading to a further erosion of trust between institution and the Labour Government. According to Opposition leader David Cameron, the Covenant was 'well and truly broken',[53] while *The Independent* warned it was at risk of being 'scandalously broken'.[54] Such statements imply culpability for the breakage, with the Government being blamed, which the Army did little to discourage. Former Defence Minister Kevan Jones MP has been one of the few Ministers happy to take on the Forces' Chiefs, who could count on allies in the media, in part, because of 'the way the political cycle was going at the time [2007–10]'.[55] The breakdown in trust between the senior commanders and the Government was reflected by public skirmishes over Chiefs' lifestyles, with Jones drawing attention to their staffed grace-and-favour accommodation. However, politicians are often most critical about military incursion into politics when they judge senior commanders to be unhelpful.

After migration, the Military Covenant continued to license senior commanders' criticism of the Labour government, while simultaneously, the official Opposition promoted the concept. Even without active collusion between them, the Conservatives and the Army appeared to share a common cause, and one that was deeply critical of Labour. This challenged the civil–military norm, affirmed in military doctrine, that the Army is 'apolitical'.[56] Richard Dannatt's appointment as an adviser to David Cameron in October 2009, described as a 'very major error of judgment' by a former head of the Civil Service,[57] added to a cumulative impression that he was anti-Labour. Indeed, for one former Labour defence minister, it confirmed that the former CGS was a 'Tory stooge'.[58] Dannatt denies this, but concedes that his 'credibility was dealt a huge blow' not least because of the premature announcement of his acceptance of the advisory

51 Strachan, 1997.

52 Robathan, interview.

53 BBC News, 4 August 2008.

54 *The Independent*, 28 February 2009, p. 44.

55 Jones, interview.

56 ADP, 2008a, paragraph 24.

57 Public Administration Select Committee, HC 994–1, 2009–10, Evidence, Lord Turnbull Q21.

58 Kevan Jones MP, interview.

role.[59] 'My motivation was and is to act in the best interests of the defence of the realm and of the Armed Forces'.[60] In the United States, civil–military relations analysis has focused on the overt Republican affiliation of the military, particularly during the Clinton era,[61] but so far there has been comparatively little academic examination of the possible compromise of the British Army's traditional party political neutrality that was engendered by the Covenant.

In connection with the Military Covenant, the Army can be regarded as a moral entrepreneur, first by its codification and then by its promotion of the concept to the civilian sphere. Having been involved in writing doctrine in the 1990s, it can be assumed that Dannatt may have been more familiar with the concept than his predecessor, Sir Mike Jackson. With other senior commanders, notably the Adjutant-General, invoking the Covenant in connection with people issues shortly after the Dannatt interview in October 2006, the inference is that the concept's deployment was a deliberate tactic by the Army – and one that was politically effective.[62] The Military Covenant allowed the Army to out-spin the Labour government.[63] Calling on the nation to uphold the Military Covenant elevates what boils down to a demand for more government money by a branch of the public sector into a moral imperative based on soldiers' service and sacrifice. In addition, seemingly unaware of the evolving nature of Army doctrine, the concept's civilian advocates all championed the Covenant in its original form.[64] However, it was this formulation, rather than the revised version that superseded it in 2005, that appeared on the Army's official website. The Covenant's entrenchment in the civilian sphere was contingent but its deployment by the Army – after years of neglect – was not.

With the civilian sphere coming to focus on just one aspect of the lattice of mutual obligation codified by the Covenant – that between the government (rather than the nation) and soldiers, the concept's civilian champions overlooked the Army's responsibilities. Kevan Jones MP says: 'Let's take one issue; housing. What were the Generals doing all those years? It was a low priority for them. They weren't innocent bystanders, they were in charge'.[65] Former Armed Forces Minister Adam Ingram concurs: 'Interesting these people have now got all these opinions about what went wrong. Were they never in control? Were they never part of the decision-making process?'[66] Had Dannatt been sacked after his *Mail* interview as Jones wanted, the Covenant might not have succeeded: that he stayed in post licensed further challenges by him and other senior commanders both to the Labour government and to existing civil–military norms. As Feaver suggests,

59 Dannatt, interview.

60 Ibid.

61 E.g. Feaver, Bacevich.

62 Sir Freddie Viggers, *Times* interview.

63 Chapter 5.

64 Including the Conservative Party, Charles Haddon-Cave QC, Mr Justice Blake.

65 Jones, interview.

66 Ingram, interview.

military obedience depends on 'calculations of whether civilians will catch and punish misbehaviour'.[67]

Although the Covenant is a tri-partite agreement, the inherent obligations of soldiers and the Army have rarely been considered. Uniquely, Forster examines this dimension of the Covenant – 'the keystone of Britain's civil–military compact' – in the context of military governance and the chain of command.[68] He suggests that senior commanders are unable to defend the Army's professional space and have fought futile battles against changing social *mores*:

> Clinging to unreflective claims that 'operational effectiveness will be damaged', waving this notion around like a lucky rabbit's foot to try to ward off the evil spirit of change.[69]

Forster's first article on the Covenant was published in *International Affairs*: the Army banned it from appearing in the *British Army Review* concerned it could prejudice the Jorge Mendonca trial. A former Army officer, Forster suggests that commanders are not used to dealing with criticism: in addition, the promotion process, reliant on the role of the Senior Reporting Officer, inhibits debate.

> It does limit people's ability to speak without fear or favour and it means that senior commanders are surrounded by people who are less willing to challenge their views, even in private, than in almost any other walk of life ... Everything is run through the chain of command, noises are silenced and marginalized, criticism and thinking outside the box is not welcomed.[70]

As the concept's first analyst, Forster initially focused the military context. Subsequent analysis has tended to overlook the 'military' aspect of the Military Covenant.

Following its migration, the Military Covenant became a shorthand term to convey the demands faced by soldiers involved in two concurrent operations that confounded existing Defence Planning Assumptions. The decision to deploy British forces in both Iraq and Afghanistan in 2006 increased the pressure on an Army that had been underfunded, undermanned and experiencing regular breaches of harmony guidelines since the 1998 Strategic Defence Review. The Covenant introduced a moral dimension to any shortfall or inadequacy in connection with soldiers' welfare. Consequently, it was far more damaging to those believed to be responsible, particularly when civilian actors subsequently extended the scope of the Covenant to include operational issues. The fractured Covenant became totemic of the absence or inadequacy of Force protection measures available to soldiers in theatre, for which the Government was blamed by Opposition MPs,

---

67 Feaver, 2003.
68 Forster, 2006, pp. 1043–1057.
69 Ibid.
70 Ibid.

coroners and the media, as well as by soldiers themselves. This is by far the most emotive issue associated with the post-migration concept, with its implication that uncaring Labour administrations had the blood of Service personnel on their hands.

As the Army promoted the Military Covenant outside the military sphere, the question about the Army's responsibility for the 'imbalanced' or 'broken' state of the Covenant in 2006 is rarely addressed. To date, Herrick continues, but already studies have been undertaken into the decision-making process leading to British participation in an expanded NATO mission, especially while the mission was far from accomplished in Iraq.[71] Whether Herrick compromised Telic was a major line of questioning in the Chilcot Inquiry. A number of analysts and observers are critical of British military performance in both theatres after 2003, challenging non-reflexive assumptions about British Forces being 'the best in the world' that are peddled by a frequently hagiographic media and by at least one military historian.[72] The broken Military Covenant, symbolising selfless troops short-changed by an uncaring government that starved them of resources, deflects attention from the Army's recent record that is, according to Ledwidge, one of 'losing small wars'.[73]

An alternative narrative about recent British military performance has been developing, puncturing the previous complacency and hubris. Wither questions the received wisdom about Britain's fabled global expertise in counter-insurgency, while Grey suggested that the withdrawal from Basra Palace September 2007 'marked, in the eyes of many, the nadir of this country's entire military reputation'.[74] In 2009, Betz and Cormack stated: 'The plain fact of the matter is that, at the time of writing, it seems entirely possible that Britain will suffer what amounts to a strategic defeat in both its ongoing counter-insurgency campaigns'.[75] *The Economist* said that seven years of war had 'exposed British military failings' and pointed out that the US was questioning 'Britain's commitment and military performance'.[76] A few weeks later, Marston said that the British were 'close to humiliation in Iraq' and 'almost destroyed' in Afghanistan.[77] Wikileaked cables suggest that US General Dan McNeill was 'dismayed' by the British effort in Afghanistan. He allegedly stated: 'They had made a mess of things in Helmand, their tactics were wrong and that deal London had cut on Musa Qala had failed'; without American support, 'We and Karzai agree that the British are not up to the task of securing Helmand'.[78]

71 Operations in Afghanistan, HC 554 2010–12: Clarke, 2011; Haynes et al., *The Times*, 9 June 2010.

72 E.g. Newton Dunn, 21 August 2009; Keegan, 2004.

73 Ledwidge, 2011.

74 Wither, 2009, pp. 611–635; Stephen Grey, Retreat from Basra – Learning the Lessons, 20 September 2009.

75 Betz and Cormack, 2009, pp. 319–336.

76 *The Economist*, 12 January 2009.

77 *The Economist*, 29 January 2009.

78 Jon Boone et al., 2 December 2010.

Operations Herrick and Telic share several characteristics and were beset by similar difficulties. Both suffered a paucity of reliable intelligence before deployment: the WMD fiasco in Iraq was matched by ill-founded optimism in Afghanistan. Tootal was told by a Royal Marine Colonel from PJHQ that it was not anticipated 'there being any trouble from the Taliban in Helmand'.[79] As Freedman observes: 'One might say to go into a theatre once and get surprised is unfortunate but, if it happens a second time, is a bit strange'.[80] Both missions descended into a protracted counter-insurgency, which had not been foreseen – or had been wished away – by planners. The absence of security in both theatres impeded the implementation of any comprehensive approach, which in turn was plagued by inter-Departmental dysfunction in Whitehall and disagreements between civil and military personnel on the ground. Both operations were characterised by a complicated command and control structure, endemic in multinational coalitions. The British public backed neither operation in the long-term. As Dannatt suggests, 'Afghanistan was very quickly seen as another of Tony Blair's bad ideas'.[81] Both missions lacked strategic coherence, not just because of the absence of any convincing rationale about how they fitted into Britain's ill-defined – if non-existent – grand strategy, but because the justification for, and the objectives of, both interventions were continually changing. Herrick 4 Brigade Commander Ed Butler observes there were 'multiple headquarters and multiple measures of success'.[82] King suggests Brigade Commanders defined their own objectives:

> It is noticeable that each brigade tour of Helmand has sought to define itself by a major operation: 16 Brigade 'broke in', 3 Commando Brigade retook Sangin, 12 Brigade 'mowed the grass', 52 Brigade retook Musa Qala, 16 Brigade transported the turbine to Kajaki, 3 Commando Brigade seized Nad-e-Ali and now 19 Brigade have taken Babaji.[83]

He argues that 'their actions have not contributed to – or have even jeopardized – long-term goals'. In addition, the six-monthly Brigade rotation – described as 'madness' by one Brigade Commander,[84] the effectiveness of which was a recurring question in the Chilcot Inquiry, 'led to discontinuity and short term thinking'.[85] In both missions, ambitions were scaled back; the endgame became the training of local security forces and the most facing-saving, calendar-based exit. Most damagingly, in both Iraq and Afghanistan, British Forces appeared overwhelmed, unable to deliver their self-selected objectives and were reliant on an American

---

79 Tootal, 2009, p. 26.
80 Freedman during Fry-Chilcot, 16 Dec 2009, p. 96.
81 Dannatt, interview.
82 Butler, telephone interview.
83 King, 2010b, pp. 311–332.
84 Officer M , telephone discussion.
85 *The Economist*, 29 January 2009.

surge. With so many of the shortcomings of the Iraq campaign replayed in Afghanistan, questions must be raised about whether the Army is still the 'learning institution' that Nagl had once described.[86]

The Moral Component of fighting power, of which the Military Covenant was an intrinsic part, is not a philosophical postscript to British military doctrine, but integral to operational effectiveness. It concerns the creation of the will to fight and win. According to military doctrine, individual and collective motivation, strong leadership and competent management are seen as crucial to engendering it. None of the critics of recent British military performance has condemned the war-fighting abilities of soldiers: without exception, they admire the courage, tenacity, professionalism and sheer grit shown by them. None has doubted their will to fight and win. Most would agree with one senior commander who told the Chilcot Inquiry: 'What worked? The soldiers were brilliant'.[87] In *Operations in Afghanistan*, the Defence Select Committee concluded:

> Armed Forces personnel achieved the best tactical outcomes possible in very difficult circumstances in no small measure due to the high quality and training of the troops themselves.[88]

This implies that soldiers have kept Covenant with the Army and nation in terms of service and sacrifice. However, what should not be overlooked is that some soldiers, including senior commanders, in fact broke the Covenant with their chain of command, the Army and the nation, as the Aitken Report highlights.

The Military Covenant codifies the Army's bond of 'identity, loyalty and responsibility' to the Nation. *British Defence Doctrine* argues 'military power is the ultimate instrument and expression of national power'.[89] In the early twenty-first century, Germany and Japan could well disagree: however, Britain's global status and membership of the UN Security Council depends in large part upon the country's willingness to deploy Armed Forces. An Army previously renowned for its skill at fighting war amongst the people appeared to lose its way after 2006. Neither Telic nor Herrick enhanced Britain's military reputation, but the missions might have far more profound long-term consequences. According to one observer:

> Like the misjudged military expedition Operation Musketeer, Suez, what it does is accelerate your decline. The failures in Iraq and Afghanistan will just add to the impression of Western weakness, Western decline.[90]

---

86 Nagl, 2005.

87 Shaw to Chilcot, 11 Jan 2010, p. 24.

88 Operations in Afghanistan, HC 554 2010–12, paragraph 11.

89 JDP, 2008, paragraph 122.

90 Interview, unattributable.

If expressions of national power are dependent upon the use of armed force, those involved in the decisions to deploy it have a responsibility to the nation. This responsibility includes not only conducting operations according customary and legal constraints and ensuring the utility of force, but also with a mindfulness concerning Britain's international reputation. As one senior commander says, soldiers 'wear the national shirt':[91] General Dannatt made much of the Military Covenant after 2006, reminding the nation of its obligations to soldiers. However, the Covenant's obligations between nation and Army are reciprocal.

The Army has an institutional obligation to the nation to be realistic about what is militarily achievable with the means available. *The Times* concluded that the Forces went into Helmand 'with their eyes shut and their fingers crossed'.[92] Hoping for the best and 'muddling through' reveals an institutional mindset that appears at odds with claims of professionalism. The deployment of British forces to Helmand in Spring 2006 was predicated on a drawdown in Iraq to a few hundred personnel. While the Helmand Plan was being drawn up at PJHQ from early 2005, security deteriorated in Basra and throughout Iraq. Herrick 4 got underway as the Samarra mosque was bombed and, according to Britain's Senior Military Representative in Iraq, 'the possibility of descent into civil war was real'.[93] As Freedman observes, someone must have said: 'Please wait, we really just can't do this until we are absolutely sure that we can get out of Iraq'.[94] If they did, they were ignored. Former CGS Jackson said the Army could handle both 'at a considerable pinch' despite being told of the possibility of 'strategic failure' in Iraq on a visit in October 2005.[95]

The British deployment to Helmand was part of an expanded NATO mission, agreed by Tony Blair in 2004. Just as the Chilcot Inquiry attempted to shed light on 'Why Iraq? Why now?',[96] a future inquiry will probably be established in connection with Afghanistan. According to Blair: 'Actually, the suggestion that we did it came from the MoD'.[97] According to Fergusson, Blair saw it as a 'swan song' and encouraged those planning the intervention to 'think big'.[98] The MoD refused to release Chiefs of Staff Committee minutes to the Defence Select Committee, which wanted better to understand the decision-making process.[99] An argument has emerged that Herrick was promoted to redeem a less than glorious Telic. The Chilcot Inquiry explored whether the British went into Helmand to try and recover

91 Sharpe, interview.
92 Haynes, et al., *The Times*, 9 June 2010.
93 Houghton, Chilcot, 5 Jan 2010, p. 9.
94 Freedman to Fry, Chilcot, p. 96.
95 Jackson, Chilcot, 28 July 2010, pp. 61–67.
96 Lyne, Chilcot.
97 Blair, Chilcot, 21 Jan 2011, p. 224.
98 Fergusson, 2009, p. 201.
99 Operations in Afghanistan, HC 554 2010–12, paragraph 15.

some status in the eyes of the Americans after Iraq. A report published, but not endorsed by the MoD, has stated that Herrick would indeed compensate for Telic:

> Any damage done by the premature UK disengagement from Iraq was partially counter-balanced [by] the strategic kudos gained from transfer of capability to Afghanistan, and the effective strategic messaging that accompanied this transfer.[100]

Fry concedes that 'there was also a view within the British Army that they could have more success in Afghanistan than they could have in Iraq'.[101] Butler attributes Britain's participation to its NATO membership: 'We had to walk the walk rather than just talk it'.[102] In addition, following the 2004 wholesale reorganisation of the Army, with regiments such as the Black Watch axed, along with the drawdown in Northern Ireland, certain senior commanders could have seen the double deployment as an insurance policy against further cuts. Tootal is sceptical: 'The military does not decide government policy. You're according the Army a level of influence they haven't had since the Parliament of 1659'.[103] However, Cowper-Coles claimed the CGS Dannatt told him: 'Use them or lose them'.[104] The deployment of Forces is ultimately a political decision. As Dannatt stated: 'I don't think we get involved in operations and wars for the convenience of the Army. We do what the government of the day requires us to do'.[105] If the Army promoted an intervention in Afghanistan out of institutional self-interest rather than in the national interest, it betrayed the Military Covenant.

How far the double deployment compromised the success of both missions has yet to be established. Chief of Joint Operations, Air Chief Marshal Sir Glenn Torpy, testified that the force structure in Afghanistan was shaped to make both theatres manageable: 'So we had constrained the size of it'. However, he did not concede that the Afghanistan campaign sucked 'resources, commitment, time' away from the Iraq campaign.[106] Divisional Commanders of MND-SE, including Shirreff and Cooper, contradict him.[107] Dannatt argued that the Government's decision to deploy in Afghanistan 'constrained' military action in Iraq.[108] Messenger, who returned from Helmand in April 2009, told the Defence Committee:

---

100 Brown, 2011, paragraph 708b.

101 Fry, Chilcot p. 96.

102 Butler, interview.

103 Tootal, interview.

104 Cowper-Coles, 23 December 2010.

105 Dannatt, Chilcot, p. 94.

106 Sir John Chilcot, Torpy to Chilcot, 18 Jan 2011, p. 86.

107 'Clearly Afghanistan had a major impact on our ability to get resources'. Shirreff, Chilcot, 11 Jan 2010, p. 11. 'We can't pretend, for example, that Afghanistan didn't have an impact on force levels in Iraq'. Cooper, Chilcot, 15 Dec 2009, pp. 80–81.

108 Dannatt, Chilcot, p. 91.

> At the time insufficient resources were being allocated to the challenge in Southern Afghanistan. I commanded a brigade, alongside an Afghan brigade commander, that was stretched and not able to go to certain key areas where we knew we would ultimately have to go to secure the population.[109]

Planning for the 2006 deployment to Afghanistan was predicated on the assumption that the Armed Forces would be, according to one MP, 'sweeping up in Iraq with a couple of guys left behind to turn the lights off'.[110] Fry admitted to Chilcot that the sequencing between Iraq and Afghanistan was 'flawed', while Dannatt conceded that the Afghanistan decision should perhaps have been revisited: 'One accepted it was a policy decision and we got on with it. Maybe that was an error'.[111] Although senior commanders are not unanimous, it can be inferred that the double deployment had an adverse impact on both operations.

The Armed Forces are an instrument of policy and therefore the ultimate responsibility for any deployment rests with politicians. However, the role of senior commanders is to give ministers military advice and, in clichéd terms, to tell truth unto power. Defence Secretary Reid appears unique in questioning the wisdom of the deployment to Helmand while Iraq was proving such a challenge.[112] During the Commons debate following his announcement about the forthcoming operation in Afghanistan, Iraq was mentioned once.[113] Torpy told him in September 2005 that 'our plans for Afghanistan are deliverable even if events slow down our Iraq disengagement':[114] Reid also received written assurances from CDS Walker that both campaigns were deliverable.[115] Although Reid's credibility with the public was tarnished by a misconstrued remark about soldiers leaving Afghanistan 'without a shot being fired', he sought repeated assurances from the military about the nature and viability of the Afghan mission.

> … All I can say is that I took the best military advice, and when we sent that number of troops in, [to Helmand] that is what the Chiefs said was necessary and I procured the funding for it.[116]

Reid delayed the mission for three months until he received assurances it was 'do-able' and that the three conditions he specified had been fulfilled.

---

109 Major-General Gordon Messenger, Operations in Afghanistan, HC 554 2010–12, Q3 26 October 2010.

110 Richard Bacon MP, Q113, uncorrected evidence, HC 43, 2006–07.

111 Fry, Chilcot, p. 93; Dannatt, Chilcot, p. 85.

112 Rt Hon Lord Reid, ibid. Q.41.

113 HC Deb, 26 January 2006, c.1529–1549.

114 Reid Operations in Afghanistan, HC 554 2010–12 Q 450.

115 Reid, Chilcot, 3 Feb 2010, p. 59.

116 Ibid. p. 63.

With British forces in Helmand involved in the fiercest fighting since the Korean War, the scope of the Military Covenant was extended to cover operational matters. The 'broken' Covenant become a trope for the inadequacy of material support, particularly in connection with body armour, armoured vehicles and helicopters – all primarily linked to Force protection in the public mind. With the media and the Opposition parties framing the issue in terms of the Covenant, the public perceived the lack of Force protection measures in terms of a failure of moral responsibility by the Government which was choosing not to honour its obligations to Service personnel, whose sacrifice was all too apparent in Helmand. The Army did not discourage this view. Consequently, although the Military Covenant was not a conscious diversionary tactic deployed by the Army to deflect attention from the conduct of military operations, it can be inferred that civilian focus on the Covenant, and Ministerial culpability for its breakage, was not wholly unwelcomed by the military.

Few in the civilian sphere have questioned the Armed Forces' responsibilities and performance that resulted in a perceived shortage of helicopters and armoured vehicles. This is a highly emotive issue. Snatch Land Rovers became known to the public as 'coffins on wheels': one SAS commander accused the Government of having 'blood on its hands' for ignoring warnings about the vehicles, in which 37 service personnel were killed between 2005–09.[117] Colonel Stuart Tootal blames the death of one of his soldiers on the shortage of helicopters,[118] while the issue reached a crescendo in the media with the death of Colonel Rupert Thorneloe and seven others in 24 hours in July 2009, following the disclosure that he had too complained about a helicopter shortage in 2008.[119] The theme of soldiers' sacrifice and government parsimony, perpetuated by the media, Opposition parties and coroners, is encapsulated by Hastings:

> The Armed Forces' 'can do' spirit was abused by the government in order to embark on ambitious operations in Afghanistan with shoestring resources.[120]

Coroners Walker and Masters, who conducted the inquests into the deaths of Service personnel, reinforced civilian society's perception of Ministerial culpability:

> To send soldiers into a combat zone without basic equipment is unforgivable, inexcusable and a breach of trust between the soldiers and those who govern them.[121]

117 [Major Sebastian Morley] Harding, *Daily Telegraph*, 7 March 2009.
118 Tootal, interview.
119 Batty, 31 October 2009.
120 Hastings, 28 August 2007.
121 Andrew Walker, James Phillipson Inquest, 15 February 2008.

The delay to inquests meant that coronial investigations into the deaths of soldiers in Iraq, often three or four years earlier, coincided with reports of military fatalities in Helmand, adding to the public perception of needless sacrifice and Government indifference.

If the Covenant were broken in Helmand, it can in part be attributed to the nature of the operation, which was very different from the one that had been planned throughout 2005 at Permanent Joint Headquarters. According to Reid, pre-deployment directives from the CDS and Joint Command contained 'very little intent' concerning the use of military force.[122] Former Brigadier Philip Mostyn was charged with producing the draft of a letter of condolence to bereaved families of soldiers killed in Helmand, with one paragraph giving reasons justifying their sacrifice. His draft paragraph was rejected by the MoD, who came back to him with a 'three-page essay', which reflects the incoherence of the objectives.[123] Within weeks of their arrival in Helmand, troops were involved in combat. The operation, in which 'lozenges' of security were supposed to be created by British forces to allow reconstruction in the Lashkar Gar district, swiftly changed in the face of attack by local Taliban. Gurkhas earmarked for guard duty were pressed into action and soldiers were bombarded in Now Zad – renamed Apocalypse Now Zad. A controversial decision by Brigade Commander Ed Butler dispersed soldiers to isolated 'platoon houses' – or Forward Operating Bases – throughout the province, where they quickly became involved in what was described as a series of twenty-first century Rorke's Drifts. Reid stated that the change in the mission's strategic nature necessitated other changes.[124]

> Let us assume that the commanders on the spot got it right; but it was an operational decision that changed the strategic nature of the mission, and when you change the nature of a mission, there is an obligation to change all sorts of things such as the force configuration, the resources and so on.[125]

The Defence Committee believes the Government should have been consulted over such a strategic shift:

> The Committee considers it unlikely that this fundamental change was put to Ministers. The Committee believes that, as the change put the lives of Armed Forces personnel at much greater risk, it should have gone to Cabinet for endorsement.[126]

122 Reid, *Operations in Afghanistan*, HC 554 2010–12 Q415.
123 Mostyn, telephone discussion.
124 Reid, *Operations in Afghanistan*, HC 554 2010–12 Q415.
125 Ibid.
126 Ibid.

The question of senior commanders' constitutional responsibilities in this matter has not yet been examined in depth. No senior commander or civil servant has admitted that prosecuting two missions was over-ambitious, or that there was consequent adverse impact on both, or that soldiers were overstretched. None resigned over the military advice given to Ministers.

The Military Covenant, which states that soldiers must be 'sustained and provided for' by the nation, will be broken by any government which makes a policy choice not properly to finance the Armed Forces, particularly at a time of conflict. After 1997, the Labour Government poured money into health, education and welfare, while leaving defence comparatively underfunded, with a declining share of GDP during a time of combat operations. Analysts, including Chalmers, suggest that the SDR was never properly funded, with a shortfall of at least £1bn a year in the defence budget after 1998. Government parsimony appeared confirmed with the Chief of Defence Logistics being given the goal of reducing costs by 20 per cent between 2000–05.[127] However, the Armed Forces are funded to prepare for war, not to fight war, money for which comes out of the Treasury Reserve. A Treasury PUS stated:

> When a war is in prospect, the narrow Treasury view that public spending is a bad thing tends to be put to one side and you get behind the national interest and you start signing the cheques.[128]

If the Treasury did indeed provide for soldiers by signing enough cheques, the money seems not to have been used to best effect. GOC-SE Shirreff said it 'beggars belief' that the British were without UAVs in southern Iraq in 2006.[129] According to Holmes: 'Soldiers in Iraq were especially galled by the fact that "the sun is hot, the rain is wet, and comms are shit"'; and by 'the sense that effective communications might be considered in some way an optional extra in a combat zone'.[130] Bishop catalogues equipment failure in Afghanistan, including Vectors whose 'wheels just kept falling off, literally' and Mastiffs out of action due to a shortage of suspension springs. A batch of faulty ammunition for .50-cal machine guns had to be dumped, with a replacement bought from Canadians for £30,000[131] and soldiers from the Royal Irish abandoned Musa Qala in local 'Jingly' trucks.[132]

The issue of Force protection measures became especially emotive among the public, with a shortage of helicopters and the use of Snatch Land Rovers perceived to cause needless sacrifice. Responsibility for the Covenant's advance to the frontline lies with politicians and the media: Snatch continues to be regularly

127 Haddon-Cave, 2009.
128 Macpherson, Chilcot, 22 Jan 2010.
129 Shirreff, Chilcot, p. 36.
130 Holmes, 2007, p. 344.
131 Bishop, 2007, p. 150.
132 Fergusson, 2007, p. 358.

cited by Conservative MPs as evidence of a breach of the Military Covenant by the Labour Government.[133] Blaming the Government for breaking the Covenant in connection with the highly-charged issue of Force protection was a diversion from two truths: the Armed Forces themselves had supported if not actively pushed for the double deployment and, as far as funding and equipping the Forces' in theatre was concerned, the Treasury had not stinted and the Urgent Operational Requirement process, although imperfect in terms of inventory tracking, steadily improved.[134]

The long-term utility of using a moral prism through which to view the issue of defence acquisition is questionable. The recommendations of the 2009 Gray Report, among them five-yearly Defence Reviews, concerned the concrete and practical rather than the abstract and moral. While acknowledging each Service's 'moral obligation' to its personnel, Gray stated 'single-Service gaming or non-cooperation must be curtailed' and advocated a legal requirement be placed on the Defence PUS as Accounting Officer to balance the defence equipment budget.[135] Such recommendations might avert the sluggishness that has frequently surrounded procurement in the decade under review, typified by lack of night vision in Chinooks going into Afghanistan in 2001, a shortfall which had compromised the Bravo Two Zero raid in Iraq a decade earlier. Combat operations highlighted longstanding problems in the acquisition process. A Deputy CDS (Capability) stated that British Forces went to Iraq with Cold War capability. He asserts: 'Some of the helicopters which were in the inventory which had been purchased against Cold War assumptions just simply weren't man enough for the job'.[136] In 2004, a National Audit Office report said that one-third of the UORs for Iraq were to fill a previously identified equipment gap.

The responsibilities and obligations of all three Services in equipment procurement decisions have been rarely assessed by media commentators and opinion-formers who castigated the government for the shortfall in Force protection measures. In 2002, after the MoD tried to smuggle through £1.1bn of extra funds to its budget during a change to accounting procedures, Chancellor Gordon Brown guillotined the Department's Spending Review assessment: this would prove to be politically disastrous for him as Prime Minister, as he was accused of cutting £1.4bn from the helicopter budget.[137] From 2006, every problem to do with helicopters – including the debacle of eight Chinooks idling away in a hanger – was blamed by the media on Government penny-pinching. However, as a former Deputy Chief of Defence Staff (Equipment Capability) told Chilcot: 'You can't buy every toy in the shop'. He added:

---

133 E.g. Andrew Bridgen MP, HC Deb, 16 February 2011, Col 1027.

134 National Audit Office, *Support to High Intensity Operations*, 2008–09, HC 508, paragraph 15.

135 Gray, 2009, p. 29, p. 116, p. 29 and pp. 34–35.

136 Fulton, Chilcot, p. 9 and p. 58.

137 E.g. George Pascoe-Watson and Andy Crick, Gord Blamed For Chopping Choppers, *The Sun*, 15 July 2009.

> If the Treasury makes a settlement of certain level then the Ministry of Defence within those resource limits needs to determine how it wishes to spend the money. The Treasury didn't say, 'You can't buy helicopters'. The Treasury said, 'Here is your sum of money' ... Clearly having enough money to buy helicopters would have been extremely useful, but the Ministry of Defence decided to spend its capital programme elsewhere.[138]

The guillotine did not fall on the budget for support helicopters such as Chinooks, but on 'the marinisation of heavy lift'.[139] The 38 per cent shortfall in helicopter capability identified by National Audit Office in 2008 became an Opposition sound bite akin to the '£38 billion black hole in the defence budget', that had similarly been identified as a possibility by the NAO but was refuted by the MoD.[140] While the Labour government should be held to account over for its long-term underfunding of defence, the perceived helicopter shortage, along with the continuing use of Snatch, became emblematic of ministerial indifference to soldiers rather than of senior commanders' unrealistic ambitions given the limited nature of the resources that the elected government was prepared to give them.

'The perfect antidote to the hybrid warrior',[141] helicopters have been central to military operations since Vietnam. If procurement had harmonised with 1998 SDR's policy direction towards expeditionary warfare, helicopters would have been more central to it. There was pressure on the helicopter fleet before the double deployment: in October 2005, CGS Jackson told CDS Walker that the Joint Helicopter Force 'was struggling to meet its tasks even with rigorous prioritization' in Iraq, where 'our support helicopter fleet is creaking badly'.[142] In the context of Afghanistan, in March 2006, the MoD recorded on behalf of Defence Secretary John Reid:

> On the matter of helicopter availability, I am reliably informed that the commanding officer of the helicopter force is content with the number of flying hours available to him for the prosecution of the mission.[143]

Former Defence Procurement Minister Lord Drayson states:

> In 2006/7 the view of the military was there was no requirement for more helicopters at the time, just a utility to having more helicopters so we could meet future requirements. The military view was also that there was no requirement for a new small helicopter.[144]

---

138 Spencer, Chilcot, 26 July 2010, p. 69.
139 Fulton, Chilcot, p. 95.
140 National Audit Office, 2010, p. 5.
141 Johnstone-Burt, cited, HC 434 (2008–09), p. 6.
142 Jackson, Declassified Extracts of October 2005 Report to CDS, Chilcot.
143 Reid, Chilcot, p. 60.
144 Drayson, 2010, p. 8.

Although Drayson pushed the MoD to adopt a more 'UOR-like urgent' approach, it was not believed helicopters could be procured quickly. Credited by many Chilcot witnesses for his leadership, Drayson got the Chinook Mk2s out of the hanger, borrowed from the Americans and took over Merlins being built for Denmark. The helicopter shortage may also be one of perception: the aircraft were pooled among the ISAF forces. In addition, availability was limited by factors such as ground crew numbers.

Vice Admiral Tony Johnstone-Burt, Commander of Joint Helicopter Command (JHC) from March 2008, inherited the 'grim events of 2007' which included crashes in Iraq, suggests that 'the fixation on aircraft numbers is missing the point'.[145] Crucial to JHC is the 'four-legged stool' of components required to generate helicopter capability, which includes a 650-personnel team and a 'robust and sustainable rule of five' involving a three-month deployment, recovery, training and pre-deployment training. 'You can only do that for so many years. Then there's the other stuff that crops up: amphibious exercises or Libya'. Johnstone-Burt describes the intense challenge not only flying 'hot and high' but of landing in a brown-out blizzard of dust, under fire from RPGs, possibly at night, with light less than 10 millilux:

> For the last two seconds everybody is blind, completely blind. The whole thing is fraught with danger ... People were doing their very best, working extremely hard and getting the most out of what they had.[146]

He had to deal with 'the consequences of the media storm', surrounding helicopter provision. His words were taken out of context in a story in which he was quoted as saying there was a possibility that a Chinook full of troops would crash: he had been asked about his worst case scenario.

> Certain members of the Press were really quite determined in the way they harangued the Prime Minister. It was numbers, numbers, numbers. This wasn't helpful for the morale of our troops.[147]

Drayson observes that, in the procurement context, JHC suffered from its joint nature, lacking Service advocates. Johnstone-Burt suggests that in the operational context:

> Joint Helicopter Command is the most brilliant organization in terms of joint working. It celebrates the single service ethos, which is enriched rather than threatened and undermined.

---

145 Johnstone-Burt, telephone interview.

146 Ibid.

147 Ibid.

Johnstone-Burt observes that helicopters were used across the Coalition forces and the British have ‘worked very closely with the US Marine Corps and the US Army in sharing aviation assets’.[148] Little research has been undertaken into the helicopter issue, but when it has, this study suggests that a more nuanced picture will emerge.

The use of Snatch Land Rovers symbolised the Government’s betrayal of soldiers. The greatest threat to troops in both theatres was from IEDs, to which the lightly armoured vehicles were vulnerable. Officer D said: ‘I’ve seen the effects. To say you’re a bit uncomfortable when you get on board is a bit of an understatement’.[149] Media interviews with dead soldiers’ parents added to the public perception of needless sacrifice due to the Government’s failure properly to fund the campaigns.[150] However, this interpretation is being challenged by the evidence that is emerging, not least from testimony given to Chilcot. As Fulton said: ‘Why were Snatch deployed to theatre? Because commanders wanted it’.[151] Drayson stated:

> Feedback from operations had highlighted the growing concern over the vehicle – however no requirement had been highlighted by the chain of command to procure a vehicle to replace it. In fact the Army were keen to keep it … The push to replace Snatch or to procure a new medium weight PPV came from Ministers not the military. For example during the meeting on 27 June 2006 between SofS, MinDP and MinAF.[152]

Following the meeting, Defence Secretary Browne ordered a review into armoured vehicles: according to media reports, the review was prompted not by ministers but by senior commanders.[153] Drayson added:

> I also pressed the Military on whether there was a requirement for a small helicopter to use for reconnaissance or surveillance to reduce the need to patrol in Snatch.

For senior commanders, policy-makers and opinion-formers to hold the Labour Government responsible on moral grounds for the problems connected with defence acquisition does little to solve its inherent structural defects. Until that is done, only then will soldiers and other Forces personnel be sustained and provided for by the nation as the Covenant demands.

---

148 Ibid.

149 Officer D interview.

150 E.g. Hennessy et al., 11 July 2009.

151 Fulton, Chilcot, p. 64.

152 Drayson, 2010, pp. 3–4.

153 Drury, 27 July 2010.

Conclusion

Having framed the challenges confronting the Army in the Covenant context in 2006, the concept's inherent reciprocity demands that an assessment is made about whether the Army as an institution, as well as individual soldiers, have honoured the Covenant. With the Military Covenant between the Nation and Army being 'out of kilter', few have considered how far the Army itself was responsible for the imbalance.

In promoting the Covenant to the civilian sphere, the Army was institutionally self-serving. The concept's deployment allowed senior commanders to challenge existing civil–military relations norms in Britain, thereby breaking their Covenantal obligations of loyalty and responsibility to the nation. Although there appears to have been no institutional intent to deploy the Military Covenant as a diversionary tactic, the effect, although inadvertent, was beneficial to the Army: while the public's attention was focused on soldiers' service and sacrifice in Iraq and Afghanistan, it was deflected from the success or otherwise of those missions. Further study could well allow a less one-sided picture to emerge about the financing of the operations and provision of equipment, particularly Force protection. Although outside the scope of this study, the question must be raised as to whether, in feeding unrealistic expectations of bloodless conflict among a casualty-averse public already steeped in a culture of health and safety, the Army undermined its *raison d'etre* in the long term.

Within the Army, *Values and Standards* had institutional primacy over *Soldiering: The Military Covenant*. After migration, the civilian sphere took control of the Covenant, determining which matters came within its purview and extending its scope to include Force protection, particularly helicopters and armoured vehicles. Civilianised, the Covenant was redefined by its advocates as a bilateral compact between the Government and soldiers, a definition to which the Army appeared to accede. Claims about a Military Covenant 'broken' solely by the Government, particularly in connection with defence procurement, is as Ledwidge suggests, a twenty-first century variation of *Dolchstosslegende*.[154] With soldiers described by much of the media and many politicians as 'heroes' for simply showing up to work, 'villains' will inevitably arise. If soldiers could do no wrong after the migration of the Covenant, the Labour government could do little right, an impression the Armed Forces did little to correct. Absolving senior commanders of all responsibility might be of benefit to them, but not to the personnel under their command, or in the long term, to the Armed Forces.

154 Ledwidge, 2011.

Chapter 7
Military Covenants

Introduction

In 2009, while the British combat operation in Iraq ended officially in April, July claimed the highest number of monthly casualties in Afghanistan, with eight British soldiers being killed in 24 hours. Consequently, the welfare of military personnel continued to resonate in the civilian sphere. The public gave the Forces unprecedented moral and material support, reflected by, for example, record donations to military charities and the celebrations surrounding the inaugural Armed Forces Day on 27 June, one of the most tangible results of the 2008 National Recognition Report.[1] In November, an External Reference Group monitoring the progress of 40 commitments given in the 2008 Service Personnel Command Paper would make its first annual report on people issues, including housing and compensation.[2] Progress was being made, particularly in connection with medical care for the wounded.[3] The nation appeared to be honouring the Covenant.

With the departure of General Dannatt as Chief of the General Staff in August 2009, the Covenant lost its principal military advocate. He left a Covenant entrenched in the civilian sphere: it had been the catalyst for focusing policy-makers' attention on people issues and had added a fresh perspective on the civil–military relationship. Much had been achieved in three years: Dannatt believed the Covenant had been brought 'back into balance'.[4] At this point it would not have been too surprising if, as was expected of the outgoing Chief, the Military Covenant had done its job and its influence had peaked. However, over the next two years the concept would become further embedded, with statutory recognition given to the Armed Forces Covenant in 2011.

This chapter examines developments in public policy in connection with the Military Covenant after August 2009. Although the Covenant retained its moral force as a result of military service and sacrifice that were manifest in Afghanistan, the Army had lost control of the concept. Various civilian actors championed and re-interpreted it. The Military Covenant's metamorphosis into the Armed Forces Covenant reflects the concept's ultimate migration from the moral and military

1 MoD, 2008a.

2 MoD, 2008b.

3 One former senior commander observes that the treatment and care of the wounded in Helmand was second to none: they were far more likely to recover than if they sustained similar injuries in a car crash on the M1.

4 Dannatt, 2011, p. 262 and p. 282.

realms to the civilian and contractual, with the emergent concept's authority being derived from statute, rather than reciprocal moral obligation. As this chapter highlights, in taking the Covenant from the moral realm, the Coalition government had to avoid entanglement in the legal component; that is, any legal liability associated with possible breaches of its statutory Covenant 'contract'. While policy-makers sought to side-step the Military Covenant's open-ended demand for 'commensurate [or 'appropriate'] terms and conditions of service', they appeared unaware that it was becoming, as Forster argues, 'quasi-judicial',[5] as the judiciary quietly recognised the concept. The Covenant was a key factor in a High Court judgment, a public inquiry and an immigration tribunal ruling.

This chapter examines how policy-making in connection with people issues was guided by the principles of 'no disadvantage' and 'citizens-plus'; that is, whether the Forces' community should be given privileged access to public services by virtue of military service. Stemming from a judicial interpretation of the Covenant is the associated question of whether military service – especially combat – should be rewarded by British residency and citizenship. The value placed on military service is one aspect determining the health of the civil–military relationship.

Military Covenants

From 2007, the Brown government reacted to the Military Covenant. Although Labour acknowledged the Covenant's fundamental principle, it refused formally to acknowledge the concept. The 2010 Defence Green Paper stated that 'How a nation looks after those who serve is an important test of its moral values' and 'The way we recruit, retain, support and reward our personnel goes to the heart of the military ethos and the nation's commitment to its Armed Forces'.[6] Even if they accepted the Covenant's precept that a debt of honour is owed to those who undertake military service, policy documents, including the 2008 Command Paper and the Party's 2010 General Election manifesto, ignored its existence. Labour's silence over the Covenant was surely deliberate, if politically short-sighted.[7] Following migration, the concept had become synonymous with Forces' welfare, but Labour refused to accept the issue being framed in a moral as well as a practical context. Many interviewees for this study state that Defence Secretary Bob Ainsworth in particular deserves credit for driving through marked improvements in connection with people issues.[8] That credit might have been more forthcoming at the time if Ainsworth and other ministers had adopted the vocabulary of the Covenant. It could well be that the Government was reluctant to move onto

5 Forster, 2012b, pp. 283–300.

6 MoD, 2010a, p. 37.

7 With Ainsworth and other Labour defence secretaries declining to be interviewed, this is speculation: however, one senior commander indicated it was deliberate.

8 Including Adam Ingram, Chris Simpkins.

territory which CGS Dannatt – with whom it had a prickly relationship – had cleared and held, and onto which the Conservatives had built, with, for example, their 2008 Military Covenant Commission. The Covenant's promotion by the Tory Opposition – who continued to claim it was 'broken' – and its avoidance by the Government, reflects the concept's politicisation. In addition, it is indicative of friction within the civil–military relationship: in its campaign to improve the lot of Service personnel, the Army's tactical deployment of the Military Covenant had left the government out-manoeuvred and on the losing side of public opinion.

The focus on 'people issues' following the Covenant's migration raised an associated question concerning the status of soldiers and the wider Forces' community in society. In giving substance to the Labour Government's moral obligation to the Forces' community, the 2008 Command Paper prompted consideration of whether its members should be 'citizens-plus', automatically deserving of privileges in connection with public services. The Paper had set out 'enduring principles for the future', chief of which were first, 'special treatment where appropriate for Armed Forces personnel, for families and for veterans' and, second, that the same groups 'should not be disadvantaged as a result of service'.[9] As the former Vice Chief of the Defence Staff who initiated the policy suggests:

> The principle of 'no disadvantage' is a minimal and dispassionate measure, which is, above all, 'future-proof' for it can be played back to Ministers on every occasion hereafter when Service people appear to be getting a raw deal.[10]

Granville-Chapman anticipated what some of the Covenant's civilian advocates had not: the concept of citizens-plus and its attendant advantages, if incorporated into public policy, could become a problematic aspect of the civil–military relationship in the future. While suggesting that Service personnel injured in the line of duty should be given priority to public services, former Army Families Federation Chief Executive Julie McCarthy, an External Reference Group member, is also wary about any special treatment being given to the wider Forces' community:

> We just want the same chances. If we go too far into citizens-plus and can't justify it, as we can for injured servicemen, there's a huge danger of a backlash. I want it to be a level playing field. For example, on school admissions we're not on a level playing field, because we rarely know our posting at the time we are having to apply for a school place.[11]

In the light of policy towards veterans being framed in terms of a 'level playing field' in 2003,[12] the post-migration debate about citizens-plus and no disadvantage

9 Bob Ainsworth, MoD press release, 19 December 2009.
10 Granville-Chapman, interview and see Chapter 5.
11 McCarthy, interview.
12 Dandeker et al., 2003, p. 17.

reflects both the previous lack of urgency to implement policies relating to the welfare of the Armed Forces and the galvanising effect of the Military Covenant on policy-makers.

Preferential access to public services again raises the tangential question of the definition of a veteran, someone in Britain who has done one day's service in the Armed Forces. The motto of Forces' charity SSAFA is 'One day's service, a lifetime of support'. McCarthy says: 'What constitutes a vet? How do you give medical priority to someone who's done a day's service?' Other public sector workers seldom remain the government's responsibility once they have left their jobs. In November 2007, the Government announced that priority access to the NHS would be extended to all veterans for medical conditions stemming from service. However, with such a benefit not being extended to others who might have been injured in the course of their work within the public sector, the Government – although perhaps guilty of gesture politics – was indicating that Forces' personnel were indeed 'citizens-plus', i.e. deserving of special treatment.

The 2010 General Election underlined how the issues associated with the Military Covenant were deployed for party political advantage. As Dorman observes, apart from the timing to replace the nuclear deterrent, there was little to differentiate three main parties regarding the bigger defence picture.[13] Although 2010 was hardly a khaki election, the steady drip-feed of sacrifice in Helmand combined with a media and Opposition eager to exploit any Government misstep in connection with Forces' welfare,[14] led to defence being co-opted into campaigning. The two main parties were keen to demonstrate their concern for the Forces: who cares wins. In the first paragraph of the first page of Labour's manifesto, Gordon Brown stated:

> This General Election is fought as our troops are bravely fighting to defend the safety of the British people and the security of the world in Afghanistan. They bring great pride and credit to our country: we honour and will always support them.[15]

Given the prominence of this statement, few reading it would infer there were no votes in defence. Although almost unacknowledged by Labour, the Covenant's influence was reflected in the Party's *Defence, Veterans and National Security* paper. This confirmed a proposal to introduce an Armed Forces Charter 'to guarantee the rights' of Armed Forces personnel and their families in areas such as housing.[16] As Drewry argues, many charters can be characterised as 'soft' contracts, 'breaches of which may give rise to financial or other penalties, but

13 Dorman, 2011, pp. 77–87.
14 E.g. Gordon Brown's letter to bereaved mother Jacqui Janes.
15 Labour Party, 2010b, p. 2.
16 Labour Party, 2010a, p. 2.

seldom have legal consequences'.[17] Labour's Charter appeared to emulate the Covenant by making explicit what had hitherto been implicit, but more stridently. The Covenant had not invoked 'rights' – apart from warning soldiers they would have to forego them – instead it had stated that soldiers 'must always be able to expect' fair treatment. An expectation and a right are very different, let alone rights that would be 'enshrined in law' and apply to the entire Forces' community.

The bond between Army and nation codified in the Military Covenant lent substance to the Conservatives' Big Society agenda, their overarching theme for government. The devolution of state power, along with a revival of mutuality and community activism were key aspirations, in place of the 'moral failure of big government' during the New Labour era.[18] Echoing Burke's little platoons, Smile's Self Help and Disraeli's One Nationism, the Big Society was not an un-Tory concept, although 'Blue Labour' guru, Maurice Glasman would argue it was stealing Labour's traditional clothes.[19] Proposing the introduction of National Citizen Service for 16-year-olds – 'non-military' but sharing 'some of the same spirit of National Service'[20] – it is unsurprising that Conservatives saw the Armed Forces, with their ethos of duty and public service, as representative of the Big Society.

> Our vision of the Big Society is one in which we all come together to solve our communal problems. It is based on respect for our great national institutions, such as the Armed Forces.[21]

Analysts saw the Big Society as an attempt to 'reassert moral considerations within politics', and as reinstating moral rather than contractual, regulation.[22] The Conservative vision was one of a less state-centric, more covenantal, society.

The Conservatives pressed the Covenant into service in the election campaign, issuing a separate 27-page Armed Forces Manifesto: *A New Covenant for Our Armed Forces and their Families*. The Tories sought political advantage by fixating on a fractured Covenant, choosing to ignore how, since the concept's migration, the public had renewed its Covenant with soldiers, reflected by donations to Service charities, the turn-out at homecoming parades and at Wootton Bassett. Society's honouring of the Covenant and third sector involvement in Forces' welfare demonstrated the solidarity the Conservatives wanted to achieve with the Big Society. The Conservatives retro-fitted their interpretation of the Covenant –

17 Drewry, 2005.
18 Ibid.
19 Glasman, 2010, pp. 59–64.
20 Conservative Party, 2010b, p. 1.
21 Conservative Party, 2010a.
22 Dawson, 2012; Jordan, 2011, pp. 1–17.

fractured – into their flagship policy, which, according to some commentators was intellectually lightweight,[23] in an attempt to give it substantive ballast.

The Coalition's post-Election *Programme for Government* was the first official policy document to refer to the Military Covenant, confirming its entrenchment in Whitehall thinking. Pre-Election pledges were inevitably trimmed by the realities of power-sharing and Britain's financial crisis, with the Programme emphasising the Coalition's priority would be deficit reduction.[24] Apart from a commitment to fulfil the 'first duty of government' by safeguarding national security and supporting 'our troops', moral imperatives were dropped in favour of practical policies.[25] Despite this, working to 'rebuild' the Military Covenant was third in a small list of defence-related measures, after the renewal of Trident and a reduction of at least 25 per cent in the MoD's running costs. Among the eight measures listed to repair the Covenant was a review of the rules governing the award of medals. Given the Conservatives' stance while in Opposition, when they included within the Covenant's scope not only welfare, but also what the public perceived as Force protection measures such as helicopter provision, this was a paltry agenda for 'people'-related issues, a herald of what would come later in the year in the Strategic Defence and Security Review (SDSR). While provision would be made for extra support for veterans' mental health needs and the doubling of the operational allowance, there was only an undertaking to 'look at' the refurbishment of accommodation financed by MoD savings. The expectations stoked by the Conservatives while in Opposition concerning the repair of the 'broken' Covenant appeared unlikely to be fulfilled.

The launch of the Military Covenant Task Force convened by Prime Minister Cameron in July 2010 further underscored the Conservatives' commitment to the Covenant concept in the context of the Big Society. Chaired by Professor Hew Strachan, the Task Force was asked to develop a series of 'innovative, low-cost policy ideas to rebuild the Military Covenant'.[26] Among its terms of reference was to 'find alternatives to "Big Government" solutions and involving the private and voluntary sectors where possible'.[27] Working to a deadline of about two months, the Task Force's small team gathered data from many in the Forces' community as well as those in the third sector and representatives of overseas' Armed Forces. Chief among the recommendations was the introduction of an Armed Forces Community Covenant, an import from America to promote links between the Forces and society at a local level, and the Chief of the Defence Staff's Commendation for those organisations that had lent outstanding support to the Forces. Apart from these innovations, there was a certain amount of reinvention

---

23 Raban (2010) suggests that the Conservatives' ideal Big Society is *The Archers*' Ambridge.

24 Cabinet Office, 2010, p. 35.

25 Ibid., p. 7.

26 MoD (Strachan), 2010b, p. 37.

27 Ibid., p. 35.

of the wheel. The Task Force went over similar ground to the 2008 National Recognition Report and the Command Paper, which in turn had gone over much that had been covered by the 2003 Report of Dandeker et al., as well as policy statements such as the 2000 AFOPS. While operational matters were beyond the Task Force's remit, the final report gave the impression that once again the Covenant was, as Tipping identified, a 'portmanteau term', with each and every welfare issue related to the Forces' 'family' held up to the light for examination before being bundled into it. In the Government Response to the Task Force's recommendations, the Community Covenant was described as being 'very much in keeping with our concept of the Big Society'.[28] However, claims about repairing the Military Covenant while seeking Forces' redundancies, cutting allowances and freezing pay seemed in keeping with critics' concept of the Big Society – a 'cynical cover for public sector cuts'.[29]

The 2010 Strategic Defence and Security Review proposed to give formal recognition to an Armed Forces Covenant. The first Defence Review for 13 years, it was undertaken in the context of Britain's straitened circumstances. As Dorman observes, the state of the economy was judged to be the country's biggest threat. Although cuts to defence were less severe than feared, Cornish suggests the Government gave the impression it was 'muddling through', trying to 'make strategic and operational success of a defence budget that is inadequate to the task'.[30] With its introduction of five-yearly Reviews, the 2010 SDSR would inevitably appear an interim measure, but when compared with its predecessor, this was compounded. In 1998, the SDR had promised to 'put our people first': 13 years later the SDSR gave an undertaking to 'rebuild the Military Covenant' and to 'rebuild and formalise an Armed Forces Covenant'.[31] The SDSR's interpretation of the Military Covenant – the Armed Forces Covenant – confirmed how the concept had been moved from *Soldiering*, soldiering and soldiers. It gave equal status not only to the three Services but also to all Forces' community members. In an unattributed echo of *Soldiering*, the SDSR stated:

> The Covenant represents a promise of fair treatment, on behalf of the nation, to ensure personnel are valued and respected as individuals and that they and their families will be sustained and rewarded by commensurate terms and conditions of service.[32]

It is paradoxical that this so closely emulates Sebastian Roberts' summary paragraph in *Soldiering*: in his view the Covenant could only apply to soldiers because of

28 MoD, 2010c, p. 4.
29 *The Economist*, 10 February 2011.
30 Cornish, 2011, p. 26.
31 MoD, CM 7948, 2010, Foreword, paragraph 2.B.3.
32 Ibid., p. 3.

the unique nature of being a soldier.[33] In what could become a hostage to political fortune, especially in a time of defence cuts, the Review extended the scope of the Covenant, stating: 'We must never send our soldiers, sailors and airmen into battle without the right equipment, the right training or the right support'.[34] More than a decade on from the SDR, the SDSR shared many similarities in connection with people issues, not least the unquestioned assumption that families or dependents of Armed Forces' personnel should continue to be the government's concern. Just as the SDSR ducked the broader issue of Britain's role in the world – a 'lost opportunity' according to 68 per cent of respondents in a RUSI poll – also unexamined was the continuing paternalistic culture of the Forces, which encourages twenty-first century Penelopes, low wages and the dependence on the imperfect agencies of the state in areas such as housing. Rather than radicalism, much in the document that related to welfare matters had a sense of re-tread: one means of delivering the Covenant 'pledge' was the principle of 'no disadvantage'.

> This obligation is not simply a moral imperative, it is fundamental to our ability to recruit and retain sufficient numbers of highly motivated and capable individuals to deliver the Defence requirement.[35]

'No disadvantage' had been identified in policy documents arising from the 1998 SDR: the difference in 2010 was that it was also seen as a moral issue.

An anaemic simulacrum of the original Military Covenant, the Armed Forces Covenant and associated policy papers were eventually published in May 2011.[36] It stated:

> The first duty of Government is the defence of the Realm. Our Armed Forces fulfil that role on behalf of Government, sacrificing civilian freedoms, facing danger and, sometimes, suffering serious injury and death as a result of their duty. Families also play a vital role in supporting operational effectiveness of our Armed Forces. In return, the whole nation has a moral obligation to members of the Naval Service, the Army and the Royal Air Force, together with their families. They deserve our respect and support, and fair treatment.
>
> Those who serve in the Armed Forces, whether Regular or Reserve, those who have served in the past, or their families, should face no disadvantage compared to other citizens in the provision of public and commercial services. Special consideration is appropriate in some cases, especially for those who have given most such as the injured and the bereaved.

---

33 Roberts, interview.
34 MoD, CM 7948, 2010, p. 5.
35 Ibid., 2B6.
36 MoD, 2011a; MoD, 2011b.

> This obligation involves the whole of society: it includes voluntary and charitable bodies, private organisations, and the actions of individuals in supporting the Armed Forces. Recognising those who have performed military duty unites the country and demonstrates the value of their contribution. This has no greater expression than in upholding this Covenant.[37]

Aiming to be 'a simple and timeless statement of the moral obligation that we owe',[38] the Armed Forces Covenant upsets the moral equilibrium that was key to the original concept. The Military Covenant was drawn up in the context of soldiering with its inherent offer of sacrifice and the responsibility for taking life. It emphasised, primarily, soldiers' obligations. According to the Armed Forces Covenant, all Service personnel face danger, as well as 'sacrificing' – rather than foregoing – some civilian freedoms. Consequently, the entirety of the Forces' community should 'face no disadvantage compared to [*sic*] other citizens in the provision of public and commercial services'. This raises the question of not only which commercial services but also of disadvantage in comparison with whom. The obligations of Forces' community members, including, for example, a drug-addicted 'veteran' who has completed one day's training are overlooked, but this veteran 'deserves' the respect and support of civilian society.

The Armed Forces Covenant codifies a rose-tinted view of contemporary civil–military relations: it implies that civilian society is content with the premise that all Service personnel are deserving of special recognition. 'Recognising those who have performed military duty unites the country' is misleading: the Covenant's migration was caused, in part, by the national disunity over the Iraq intervention and subsequent efforts by CGS Dannatt to rally support for soldiers among an apathetic public. Like other policy documents setting out civilian interpretations of the post-migration Covenant, inherent in the Armed Forces Covenant are implied notions of military moral superiority. Service personnel are, by implication, deemed to be 'citizens-plus' or, in a phrase associated with Reservists, 'twice the citizen', all automatically deserving of civilian society's respect. For example, the Conservatives' Armed Forces manifesto of 2010 states: 'The sacrifices being made continually by all our Armed Forces, however, show that their side of this covenant is without doubt being upheld throughout'.[39] At minimum, this statement involves air-brushing from recent history various 'war crimes' trials, the Baha Mousa controversy and the ignominious detention of British sailors by Iranian forces in 2009. Since migration, the public had voluntarily upheld the Military Covenant through conduits such as Help for Heroes. In the long-term it might be less happy being exhorted by the State unconditionally to support Service personnel, especially those who have broken the reciprocal terms of the Covenant as laid down in Army doctrine, by, for example, abusing Afghan detainees.

37 MoD, 2011a, p. 1.

38 Liam Fox MP, HC deb 16 May 2011, c25.

39 Conservative Party, 2010a, p. 11.

Covenant-related policy documents imply that Service personnel are a privileged caste by virtue of their profession. While opinion polls consistently suggest that the general public has a favourable view of the Armed Forces, there has been little analysis of whether it wants all Service personnel to be regarded or treated as 'citizens-plus'. The 2010 National Security Strategy – predicated on 'risk' rather than threat, according to Edmunds[40] – identifies that 'tier one' risks to Britain are international terrorism, cyber attack, a natural hazard and 'an international military crisis, drawing in the UK and its allies'.[41] Given such risks, Servicemen and women are not wholly responsible for national defence; however, according to the Armed Forces Covenant, the nation owes them a 'moral obligation'. Today, Britain enters into 'small wars' involving volunteer professionals, which, even if unpopular, are still 'wars of choice'. They do not concern the defence of the realm against an existential threat, so much as the defence of Britain's place in the world. This depends, in large part, on a readiness to deploy the Armed Forces, something that was especially apparent during the Blair era.

Written for the civilian rather than military sphere, the Armed Forces Covenant was the culmination of the post-migration process of civilianising the Military Covenant. Although the Army might have lost control of the concept, its deployment had not only forced civilian policy-makers to address the Army's concerns, but also more frequently to affirm, publically, their support for soldiers and the Armed Forces. Even if perfunctory or platitudinous, statements that extol the military surely benefit the Army as an institution. In the United States' context, analysts have examined the phenomenon of military moral superiority,[42] in which personnel regard themselves as more worthy than civilians, a notion bolstered by politicians who, as Bacevich describes, view military institutions 'as the chief repositories of civic virtue'.[43] Although British Service personnel are polled annually, their attitudes towards the civilian sphere are not canvassed; consequently whether a similar sense of military moral superiority exists on this side of the Atlantic is unknown.[44] However, as this study has examined, the Moral Component not only consciously set the Army apart from civilian society in terms of values, but also was engendered by an institutional mindset that implied a moral superiority to 1990s civilian society.[45] Although the codification of Army's values and standards was necessitated by military imperatives, it could not but imply that an unfavourable judgment had been passed on the civilian sphere. Like their American counterparts, British politicians now routinely pay tribute to military virtue.[46] While this is to be expected, especially during a combat operation, the

40 Edmunds, 2012, pp. 265–282.

41 HMSO, 2010, p. 27.

42 E.g. Ricks, 1998; Bacevich, 2005.

43 Bacevich, p. 32.

44 The annual Armed Forces' Continuous Attitude Survey.

45 Chapter 3.

46 E.g. 'The bravest and the best', Tony Blair; 'The best of British', David Cameron.

Armed Forces Covenant has translated such sentiment into policy. Paradoxically, civilianisation represents the military's moral victory.

In giving formal recognition to the trust-based Military Covenant in public policy, the concept was transformed from the moral and military towards the civilian and transactional. Crucially, unlike the Military Covenants codified in military doctrine, the Armed Forces Covenant offered soldiers and other Service personnel neither 'commensurate' nor 'appropriate' terms and conditions of service. It can be inferred that Coalition policy-makers were keen to avoid 'writing blank cheques'.[47] However, it was precisely to avoid legal notions of liability that the term 'covenant' was deliberately chosen by those involved in developing the Moral Component of Army doctrine. To minimise the legal minefield it had stumbled into, the Government had to rewrite Paragraph 103 of *Soldiering*, erasing any reference to material recompense – which after all is in the government's gift – in exchange for military service and sacrifice, instead stating that the 'whole nation' and the 'whole of society' should offer the moral rewards of respect, support and fair treatment. Prime Minister Cameron unwittingly revealed the Government's attitude in an article in the *Sunday Telegraph*:

> The truth is there is no hope of properly repaying our Armed Forces for the work they do, either through money or resources or gratitude. But we – as a country and as a government – must always do our best to show our thanks and our pride in them; just as they have always done their best by us.[48]

Armed Forces personnel might suggest there is every hope of properly repaying them, not least by upholding the Military Covenant written for Army doctrine; this stipulates fair treatment, value and respect, together with commensurate or appropriate terms and conditions of service. In addition, the prescriptive exhortation that the Covenant's obligation 'involves the whole of society' implies a refusal to acknowledge the whole of society's ambiguous attitude to the Armed Forces – a combination of respect, admiration, indifference, ignorance and an unwillingness to pay more tax to support them.

The process of statutory recognition, involving the rewriting of the Military Covenant, emphasised the civilian appropriation of the concept. In June 2010, aboard HMS *Ark Royal* the Prime Minister announced that a 'rewritten' Covenant would be 'written into the law of the land'.[49] The announcement to sailors on a warship underlined how far the Covenant had migrated from Army doctrine. Statutory recognition – to be realised in a clause in the Armed Forces bill – attracted a great deal of party political wrangling in its passage through Parliament between December 2010 and November 2011. Having previously shunned the concept, in its Opposition Day debate of February 2011, Labour now supported 'a

47 Roberts, Ritchie, interviews.

48 Cameron, 19 June 2010.

49 Cameron, 24 June 2010.

legally-binding Military Covenant which defines the principles that should guide Government action on all aspects of defence policy'.[50] Debates on the proposal gave MPs the opportunity for partisan critiques of the management of defence[51] Labour 'did not adequately equip our troops for the most hostile of conflicts' while the Coalition's change to Forces' pensions was 'taking money from Afghan war widows'.[52] With the Armed Forces Covenant not officially published until May 2011, Parliamentary debates and discussion about 'enshrining the Covenant in law' took place in the context of the Military Covenant that had been codified in Army doctrine. For example, in the Armed Forces bill's Second Reading, MPs had quoted the 2000 and 2005 summary paragraphs from military doctrine.[53] However, according to these two versions of the Covenant, soldiers and their families should be able to expect either 'commensurate' or 'appropriate' terms and conditions of service.

The process of giving the Military Covenant statutory recognition reveals an attempt by policy-makers to reconcile the irreconcilable: the military Moral Component with the civilian legal component. Not only was the Covenant radically rewritten, but the government appeared to want to retreat from its commitment to give the concept legal recognition, underscored when a defence Minister in a Select Committee hearing in February 2011 described the Covenant as a 'conceptual, philosophical statement'.[54] Instead, policy-makers proposed that an annual 'Armed Forces Covenant Report' should be presented to Parliament. Essentially, this proposal was a diminished 2008 Command Paper annual report that did not have the benefit of independent scrutiny by an External Reference Group. It was a somewhat anti-climactic ending to relentless moral grandstanding by the Conservatives since 2007 on the issues they considered relating to a Military Covenant, 'shattered' by New Labour,[55] and whose scope had included everything from helicopter numbers to defence acquisition. The Report proposal drew widespread criticism, not least from the Royal British Legion, which sent an email to all MPs:

> We do not understand why the Government is now claiming that the commitment to produce an 'Armed Forces Covenant Report' is somehow the same thing as enshrining the Military Covenant in law. It is not the same thing at all.[56]

---

50 Jim Murphy MP, 16 February 2011, HC Deb c1026.

51 House of Commons, Opposition Day Debate, 16 Feb 2011.

52 Jack Lopresti MP, 10 Jan 2011 HC Deb c101, Jim Murphy MP HC Deb c63.

53 Charlie Elphicke MP 10 January 2011; HC deb c109; Christopher Pincher MP 10 January 2011 HC deb c105.

54 *Official Report, Armed Forces Public Bill Committee*, HC 779, 2010–12, Evidence, Andrew Robathan MP, 10 February 2011; c.22.

55 Fox, 2010, speech, Birmingham.

56 Cited by Jim Murphy MP, 16 Feb 2011 HC Deb c1032.

In May 2011, the Government delayed the Armed Forces' bill following pressure from the Legion, a *News of the World* campaign and an amendment tabled by Philip Hollobone MP to 'enshrine the Military Covenant in law'.[57] Less than a fortnight later, came the announcement that the new Armed Forces Covenant and its core principles of 'no disadvantage' and special treatment 'where appropriate' would be 'enshrined in law'.[58] For almost a year, the Coalition's oscillations regarding a statutory Armed Forces Covenant suggest that, while the concept itself had become established as an inextricable part of defence policy and synonymous with people issues among policy-makers, there was anxiety that it may become justiciable. The Government wanted, unsurprisingly, the best of all possible worlds: tangible proof that it was on the side of Service personnel by extending the Covenant from a moral obligation to a statutory one, while also ensuring that any future failure to fulfil the Covenant with those personnel, their families or veterans would not involve the Government in civil litigation, resulting in compensation payments.

Among serving or former soldiers, including those interviewed for this study, there was little, if any, support for putting the Military Covenant on a statutory basis. One senior commander told the Armed Forces Committee: 'Military Service is not a contract and that is not an area where we want to go'.[59] With one exception, military interviewees were wary about anything that might involve the Army in additional litigation or could impede operational effectiveness. Sebastian Roberts says:

> One of the things I rather like about this – that I'm proud about – is that it is not legally binding, it is not a contract. Although I and others might have once thought it would be lovely if it were a legal contract, once a thing becomes enshrined in law – I don't mean this in any cheap way – it falls into the hands of lawyers. Then it becomes argued about and the very words get picked apart, not just by lawyers but by the politicians and the pressure groups. … Suddenly it becomes the stuff of employment tribunals and appeals for money. I would rather keep it human and flexible rather than it should be bound by mathematics and law. It is a human document. It is about the human component of soldiering, not the physical or intellectual. This is designed to deal with things that are essentially themselves intangible.[60]

Stuart Tootal mooted that a statutory Military Covenant should have legal teeth,[61] a suggestion which would not have found favour with the Government. As the Defence Secretary stated in May 2011:

57 Norton-Taylor, 5 May 2011.

58 Liam Fox MP, 16 May 2011, HC Deb c25.

59 Armed Forces Select Committee, HC 779, 2010–12, Evidence, General Sir William Rollo, 8 February 2011, Q232.

60 Roberts, interview.

61 Tootal, interview.

> In deciding how best to recognise the covenant in law, the Government have had to maintain a careful balance … We do not want to see the chain of command undermined or the military permanently involved in human rights cases in the European Courts.[62]

With some legal firms now specialising in compensation claims against the MoD, the Government was rightly nervous. The Armed Forces Covenant attempts to harness the moral authority of the Military Covenant, the current popularity of soldiers and the principles set down in previous policy documents, in a statement that is also 'lawyer-proof'.

The Armed Forces Covenant represents the entrenchment of Military Covenant in the civilian sphere. How far 'enshrining the Covenant in law' has a tangible positive impact on the welfare of the Forces' community beyond, for example, the policies put in place by the 2008 Command Paper, is, at the time of writing, to be determined. No sanction exists if the Armed Forces Covenant is broken, shattered or remains unrepaired. As Pyianno's post to the Army Rumour Service observed:

> The Covenant Report seemingly imposes no obligation upon the SoS other than to make a token speech once per annum, in order that his complacent backbench MPs might wave through his efforts as perfectly satisfactory, in spite of the real state of affairs.[63]

Giving legal recognition to the Covenant is another example of 'aspirational legislation' examined by Elliott.[64] Such legislation lays down requirements that are 'largely unenforceable in a legal sense'. Two recent examples are targets for ministers to attain on climate change and child poverty. As Elliott suggests: 'it does not follow that failure to achieve such targets will be actionable in court'. Aspirational legislation is part of politicians' recent tendency to legislate to be seen to be doing something: it can be nothing more than a gimmick. To date, members of the Coalition government regularly refer to the 'Military Covenant' and trumpet its legal status as an achievement. They see no irony or paradox in connection with their statements about 'enshrining the Military Covenant in law' while sacking Service personnel, reducing Service pensions and allowances, or appearing to make a policy choice to manage the further decline of defence.

While politicians wrestle with the post-migration Covenant, they appeared unaware that the concept was influencing public policy through the agency of the Judiciary. Since 2005, it has been a factor in two public inquiries including Deepcut, a High Court judgment and an Immigration Tribunal Appeal all of which embedded the concept further in civilian society, independently of the Parliamentary process. As MPs were demanding that the Covenant be repaired,

62 Liam Fox MP, 16 May 2011, HC deb c25.

63 ARRSE, 8 November 2011.

64 Elliott, 2010.

jurists outside Parliament were interpreting the concept, transforming it into the 'quasi-judicial concept' identified by Forster,[65] with perhaps more far-reaching consequences for Service personnel than giving statutory recognition to the Armed Forces Covenant. That a military concept should influence civilian law represents a small reversal of the general trend, which has seen law originating in the civilian sphere – particularly concerning human rights – encroaching ever more deeply into the military's professional space.[66]

The courts have ruled that in connection with citizenship rights, military service – particularly combat – does indeed bestow special privileges. The Military Covenant has been salient in two cases concerning immigration, one of the most contentious areas of British public policy. The Covenant adds another dimension to the national immigration debate, crystallising the delineation between the deserving and undeserving immigrant. In the decade under review in this study, the Armed Forces actively recruited Commonwealth citizens. Compared with 1,912 in 2001,[67] the number of foreign soldiers rose from 6.6 per cent in 2007 to 7.9 per cent or 7,750 in 2010.[68] This increase in numbers raised periodic alarms in the media about the Army becoming a 'foreign legion'.[69] In early 2009 the Army placed a 15 per cent cap on the number of non-UK personnel serving in three Army Corps, with an additional option to extend the limit to other regiments in the future. The measure underlined 'the importance of ensuring that the Armed Forces continue to be identified with and representative of the UK'.[70] This statement highlights a somewhat arbitrary notion of identity and representation in contemporary Britain. As Kaldor observes, narratives of identity politics depend on memory and tradition.[71] Without doubt, the Forces are identified with the nation, and also by the nation, as a key aspect of Britain's national story, particularly in connection with Empire: however, they are hardly representative of the nation. Britain's population is estimated to include some 4 million foreign citizens, 1.5m from the EU. Apart from members of the Brigade of Gurkhas, aliens – those who owe no allegiance to the Crown by reason of birth or nationality – have been prohibited from serving in the Armed Forces since the 1701 Act of Settlement. In addition, although Muslims are an estimated 4.6 per cent of the British population, they represent just 0.3 per cent of the Armed Forces.[72] With the exception of those from the Republic of Ireland, a ban has also remained in place on European Union citizens joining the Forces. With their legally sanctioned acceptance of Commonwealth rather than

65 Forster, 2012b, pp. 283–300.
66 Forster, 2006, pp. 1043–1057.
67 Lewis Moonie MP, 26 October 2001 HC c405W.
68 DASA 2010, Table 2.14.
69 Evans, 14 November 2005.
70 John Hutton MP, 2 February 2009, HC Col 34WS.
71 Kaldor, 2007, p. 7.
72 Pew Forum on Religion and Public Life, cited Maher, 2011, p. 60.

EU citizens as recruits, the Armed Forces are atypical and unrepresentative, not only of all other major employers, but of Britain itself.

A High Court judgment in 2008 concerning Gurkha veterans in which the Military Covenant was key, challenged existing immigration policy and raised the possibility that 36,000 former Gurkhas and their families could settle in Britain.[73] Although Armed Forces personnel from overseas had been granted relatively minor privileges in connection with entry to Britain since the early 1960s, until 2004, it was 'almost impossible'[74] for Gurkha and Commonwealth members of the Armed Forces to be granted settlement or British citizenship. Indeed, in 1980 when the Thatcher government was looking at changes to the nationality laws, citizenship as a reward for Crown service was discouraged:

> It is envisaged that grants of Citizenship on these grounds would be made only sparingly. It would not be right, for example, to make them simply on grounds of a period of satisfactory service; citizenship, carrying with it the right of abode in the United Kingdom, is not appropriate as a form of emolument.[75]

In October 2004, the Home Office published the new HM Forces rule, allowing Forces' personnel with at least four years' service to apply for settlement following their discharge from service. However, Gurkha veterans who had served before July 1997 were exempt from this entitlement. The subsequent controversy involved the actor Joanna Lumley, reminders of the Gurkhas' two centuries of service to Britain and 84-year-old Tulbahadur Pun VC, veteran of the Burma campaign, who was refused a visa by the British Embassy in Kathmandu because he had 'failed to demonstrate close ties to the UK'. The Gurkhas' plight led not only to a reappraisal of public policy, but provided an insight into public attitudes towards 'soldier-citizens'.

In September 2008, the Military Covenant was material to a High Court judgment (*Limbu*) following a legal challenge by a group of Gurkhas, who had served before 1997 but had been refused entry to Britain. Mr Justice Blake, familiar with the Covenant from his 2005 inquiry into Deepcut, ruled that the criteria governing the exercise of discretion by immigration officials in connection with the Gurkhas was 'irrational' and should be reviewed. He noted that the 2004 change of policy that granted settlement to those Gurkhas who had served after 1997 was said to be in recognition of the debt of gratitude the nation owed to the Gurkhas for their loyalty and commitment to the Crown through their service in the British Army. The Court 'was conscious that at the heart of military life and the sacrifices that soldiers make in the discharge of their

73 *R (on application Limbu and others) v Secretary of State for Home Dept & Others*, EWHC 2261, 2008.

74 Thorp and Woodhouse, HC SN/HA/4399 July 2009.

75 White Paper, *British Nationality Law: Outline of Proposed Legislation*, July 1980, Cmnd 7987, paragraph 69; cited Gower, SN/HA/3232, 2011.

duties is the Military Covenant'.[76] Quoting in full the summary paragraph from *Soldiering*, the Judge added:

> Rewarding long and distinguished service by the grant of residence in the country for which the service was performed would, in my judgment, be a vindication and an enhancement of this Covenant.

Despite Ministerial insistence that 'the Executive has to look at the consequences for the taxpayer, not just at the moral issues involved',[77] in May 2009, the Government aligned settlement rights of all Gurkhas, a policy reversal that was the result of moral pressure exerted by the Gurkhas' record of military service to Britain. The *Limbu* judgment not only entrenched the Military Covenant deeper in civilian discourse, but emphasised not just a reciprocal moral reward of 'respect' and 'fair treatment' that soldiers should expect, but also the material; in this instance the right of residence and ultimately citizenship. How far it sets a precedent is unknown.

Entangled as it was with the politically charged and socially divisive issue of immigration, the Gurkha controversy highlighted how military service to Britain was deemed to justify settlement privileges. Rifleman Pun's plight attracted almost 2,500 posts on the Army Rumour Service website, which galvanised public opinion through viral campaigning. Some posts distinguished between the Gurkhas and those judged less deserving, a point of view emphasised by some tabloid columnists, one of whom declared, 'British people want Gurkhas not shirkers'.[78] This underlined the value placed upon military service by not only jurists but also by the public and policy-makers, who from 2004, had deemed all non-UK citizens in the Armed Forces deserving, while signalling their particular reverence for frontline service.

The Gurkhas in particular and Commonwealth troops in general have refashioned the traditional ideal of the citizen-soldier for a contemporary British context. The citizen-soldier is a civil–military construct that historically had little salience in Britain compared with countries where compulsory military service has been a mark of citizenship. The granting of residency and, by extension, citizenship to non-British soldiers in exchange for their military service reflects the privileged cultural status of the military that has been identified in connection with the US Armed Forces.[79] However, as Krebs suggests citizen-soldiers have, in fact, been relatively rare in the recent history of the United States, with compulsory and 'relatively universal' military service lasting from the beginning of the Second World War to the end of the Korean War. Despite that, 'the citizen-soldier ideal

76 Ibid., paragraph 72.

77 Phil Woolas MP, 29 April 2009 HC deb c997.

78 Gaunt, *The Sun*, 1 May 2009.

79 Krebs, 2009, pp. 153–174.

has resonated strongly'.[80] Indeed, after the declaration of the Global War on Terror, the United States itself had to invert the conception of the citizen-soldier, with soldiers becoming citizens in exchange for their service. In February 2009 the MANVI programme (Military Accessions Vital to the National Interest) offered skilled immigrants such as doctors, who had been granted temporary work visas but not green cards, citizenship in as little as six months. In July 2002, announcing that non-citizens serving in the US Armed Forces would be immediately eligible for citizenship, President Bush declared military service 'the highest form of citizenship'.[81] Unlike the United States, Britain denies many of those who have settled within its borders the right to serve, while similarly extolling the virtues of military service and Service personnel. The implication is that other immigrants are less deserving compared with those 'soldier-citizens' who have served.

The quasi-judicial nature of the Covenant was reinforced by a second immigration case in which the concept was a material factor, together with the Human Rights Act. A veteran of Iraq from Fiji faced deportation by the Home Office in early 2011, but won a successful appeal. Former Private Epeli Uluilakeba was first injured by a roadside bomb that killed three of his fellow soldiers travelling with him in a Snatch Land Rover and then, on his return to Iraq, survived another IED. After assaulting an NCO, he was court-martialled and sentenced to a military prison before being discharged from the Army in December 2009. Consequently, the Home Office refused him leave to remain in Britain. However, at an immigration appeal, the Judge found that in accordance with Article 8 of the European Convention, Uluilakeba had established a private life in Britain, not least because of his need for healthcare resulting from his service in Iraq.

> Finally and of great significance, is the fact that the removal of this applicant would amount, in my judgment, to a breach of the Military Covenant, which I am satisfied must cover a responsibility of the State to provide physical care to service personnel for physical injuries sustained in operations and mental support for post traumatic stress disorder.[82]

The case reflects how the moral component is starting to become intertwined with the 'legal component', with, to date, uncertain consequences for the Army.[83]

The scope of the Military Covenant was extended by the Haddon-Cave Review into the loss of Nimrod XV230. The plane crashed in Afghanistan on 2 September 2006, with the loss of all 14 Servicemen on board.[84] Haddon-Cave concluded that 'XV230 was lost because of a systemic breach of the Military

80 Ibid.

81 Ibid.

82 First Tier Tribunal, Immigration and Asylum Chamber, 16 May 2011, p. 12.

83 E.g. possibly fewer Commonwealth soldiers.

84 Haddon-Cave, 2009.

Covenant brought about by significant failures on the part of all those involved'.[85] Those involved had not only included RAF personnel but civilian employees from private sector organisations including BAE Systems and QinetiQ. The Review was clear that, with the increasing civilianisation of defence support, those working in the defence sector should be regarded Covenant agents:

> The history of In-Service Support has seen a shrinking of the role of uniformed and Crown personnel and commensurately greater devolvement of responsibilities to, and reliance, on the Defence Industry. It is important, for the future, that all organisations involved in this endeavour recognise their collective responsibility to work together to ensure that the Military Covenant is never broken, as it was in the case of Nimrod XV230.[86]

The government-ordered Review was to have a significant impact on public policy, not least with the creation of the Military Aviation Authority and a promised re-examination of the links between government, the Services and the defence industry. Haddon-Cave made clear that, although the Covenant was commonly understood in the context of the care and support given to wounded Service personnel and the families who have suffered loss, its scope is far wider:

> ... And embraces the whole panoply of measures which it is appropriate the Nation should put in place and sustain for Service personnel, including adequate training, suitable and properly maintained equipment, sufficient provisions in theatre and proper support and conditions for Service personnel and their families at home ... The same rationale applies to the Navy and the Air Force.[87]

In demanding measures such as Force protection to be included within the remit of the Military Covenant, Haddon-Cave went further than any politician who had demanded that the Covenant be 'enshrined in law'.

The consequences of the recognition given to the Military Covenant by Haddon-Cave are significant. The Review described the Military Covenant as the 'modern expression' of a 'sacred and unbreakable duty of care' that is owed to the men and women of the Armed Forces.[88] However, if what is unbreakable is actually broken, as a result of the tort of negligence under common law, legal recompense can be sought. Although the exact figure remains confidential, compensation was paid to the Nimrod victims' families, with the Ministry of Defence not contesting its legal duty of care. It is unlikely that even a 'sacred' moral duty of care would have persuaded the Department's legal advisers – who would be mindful of precedent –

85 Ibid.
86 Ibid., paragraph 29.4.
87 Ibid., paragraph 29.3.
88 Ibid., paragraph 29.1.

of the need for a multi-million pound settlement.[89] In addition, with two of the personnel named in the Review placed under criminal investigation, the legal trumped the moral. As a concept with a moral rather than a legal foundation, the Covenant is reliant upon trust for its fulfilment.[90] There are few other activities where individuals cede such control and have to entrust others with their safety than when they are in an aircraft. With compensation being paid and a criminal investigation being instigated, it is not unreasonable to assume that the inclusion of considerations about the Military Covenant were superfluous to the Report. However, far from being an irrelevance, the Military Covenant was subject to another re-interpretation by Haddon-Cave, with a precedent being set for civilian contractors to be drawn into its orbit. It is understood that Charles Haddon-Cave regards the final chapter of his Review, which explored the Covenant concept and deliberately extended its scope, as the most important.[91] His radical interpretation of the Covenant was a message to both the public and private sector, which left little doubt that he expected the Armed Forces to be properly equipped and trained to an irreducible minimum in the future.[92]

Its civilian advocates deployed the original codification of the Military Covenant. Election manifestos and other political literature, as well as legal judgments and government-ordered Reviews, whether Blake or Haddon-Cave, all relied on the formulation laid down in Army doctrine in 2000, which was not only less equivocal than its successors but was historically inaccurate.[93] Policy-makers appeared unaware that the Military Covenant was not set in stone, but a living document and subject to amendment: it was updated by the military, first in 2005, and then in 2010. The Haddon-Cave Review, like the *Limbu* judgment, invokes the 2000 Covenant, quoting the summary paragraph in full in a footnote. The Review states that the duty of care owed to Service personnel who are ‘necessarily called upon to make substantial personal sacrifices, including the ultimate sacrifice’, dates back to 1593. The Conservative Party’s Military Covenant Commission cited the same date. Whether an Act passed by Parliament more than 400 years ago, the implementation of which was the devolved duty of the imperfect apparatus of local parishes, should be any basis for twenty-first-century public policy is open to question. That the Elizabethan state might have had self-serving rather than altruistic reasons to introduce such a policy, that the policy might rapidly have fallen into desuetude, that early seventeenth-century attitudes towards soldiers and mariners were far from favourable, has not been considered by those who invoke a statute set down almost 70 years before the embryonic form of the modern British

89 Reportedly £14m.

90 See Chapter 4.

91 Private information.

92 In April 2014, a report linked the introduction of Duty Holder Concept regulations with the Haddon Cave Review (Mark Nicol, Senior Officers who Send Troops to their Deaths Using Faulty Kit Will Be Jailed, *Mail on Sunday*, 6 April 2014).

93 Chapter 1.

Army came into existence. Just as military doctrine evolves, the 1593 statute was superseded by those of 1597 and 1601, which also promised support for soldiers and mariners. One counterfactual to consider is whether, without Sebastian Roberts' gift for writing reflected by Paragraph 103 of *Soldiering*, the Military Covenant would have appealed to, and become entrenched in, the civilian sphere.

Although the Covenant had become civilianised, the concept retained its institutional utility for the Army. In 2010, a revised Military Covenant was included in the new capstone doctrine publication, *Operations*.[94] This Covenant was similar to its 2005 predecessor, stating 'the nature of Service is inherently unequal in that Servicemen [not soldiers] may have to contribute more than they receive'.[95] In return, they 'and their families should be rewarded with appropriate terms of service'. In a departure from the two previous codifications, the Army's 2010 Covenant – obliquely – raised the issue of properly resourcing the Armed Forces. '... They [Service personnel] should be able to expect the Nation, and their commanders, to provide them with the means and ways to achieve the ends set'. However, Major-General Andrew Sharpe, who contributed to *Operations* suggests soldiers are pragmatic about this:

> You can't make a policy that says 'you will always provide the best possible kit humanly imaginable for your Army before you put soldiers in harm's way'. You can't do that, and who's to say what kit it is, because a soldier would probably say, 'Would I rather do without my body armour all together but have lots of helicopters?' Under some circumstances, absolutely not: under others, yes, absolutely.[96]

Despite being entitled the Military Covenant, it was now between 'the Nation, the Service and each individual soldier, sailor, marine and airman', reflecting that, although *Operations* was an Army Doctrine Publication, it received Joint scrutiny, with endorsement by the RAF and Royal Navy.[97] Unsurprisingly, recent military doctrine writers bring a military perspective to understanding the Military Covenant and its impact on soldiers' welfare. Sharpe explains:

> If it's a question between how many battalions can we have and how comfortable will your home be, I know where the judgment will fall. And actually your average bloke in that battalion, if he were told: 'Are you prepared to live in a grotty house, so the battalion lives on in the next round of cuts that are bound to come?' I would say the large majority of them would say, 'I don't fucking like it sir, but I'll knuckle down'.[98]

---

94 ADP, 2010, paragraph 0230.

95 Ibid.

96 Sharpe, interview.

97 Sharpe, note.

98 Ibid.

As Sharpe suggests, soldiers keep Covenant with their regiment, something that has been overlooked by many civilian advocates of the concept. Echoing its 2005 predecessor, the 2010 Military Covenant states: '… It is a covenant, not a contract'; therefore it imposes 'moral rather than legally binding obligations'.[99] Sharpe says: 'The line has always been that it's not a contract; it's binding but it's not a contract'. He stresses the principle of reciprocity inherent in the doctrinal Covenant:

> This is as much telling Servicemen who read this that they are beholden as it is reminding them that we feel that the nation is beholden back to them.[100]

In 2010, Military doctrine underscored that the Military Covenant was a moral concept based on trust and reciprocity: in short, it is a covenant.

Conclusion

Since its migration, the Military Covenant has redefined, first, soldiers' welfare and then the welfare of the entire Armed Forces' community as a national policy imperative. With their welfare being recast in moral terms, Forces personnel were also being recast as citizens-plus, deserving of special recognition and privileges, including, for some, the right of residency. However, this could have far-reaching consequences, not least the end of Gurkha service to Britain. After the financial crisis of September 2008 all policy, including defence, was framed in the context of budgetary constraints. Although by adopting the Covenant, policy-makers had endorsed and championed a moral agreement, implementing Covenant-inspired welfare policies would be subject to mundane administrative process. This was summed up by an accompanying document to the SDSR, in which the MOD stated:

> Honouring the Covenant doesn't necessarily have to mean spending large amounts of money. Ensuring that Service personnel, their families and veterans are treated fairly can often be about adapting existing policies where the particular needs of the Service community had not previously been taken account of. Many of the commitments in the Programme for Government will be led by other Government Departments and will not be reliant on defence funding being made available.

The Armed Forces Covenant, which essentially rehashed the Service Personnel Command Paper, faces exactly the same problems of cross-Departmental implementation that has troubled people issues since the 1998 Strategic Defence

99 ADP, 2010.

100 Sharpe, interview.

Review. However, with the annual Armed Forces Covenant report, the Government has created an audit trail in relation to its performance.

The Military Covenant written by Sebastian Roberts was inspirational, capturing the imagination of the nation and encouraging the public to honour their moral responsibilities to Forces personnel serving in Iraq and Afghanistan. The public came to uphold the Covenant, as General Dannatt acknowledges. However, out of partisan self-interest, the Conservatives continued to insist it was fractured. Their solution to a receding problem – public indifference to the Armed Forces – was unnecessary legislation. Seeking to put the Covenant on a statutory footing, the Coalition retreated when it threatened to be justiciable, instead, creating the lawyer-proof Armed Forces Covenant. However, the welfare of the Forces' community concerns political will and public spending choices. The necessarily anodyne Armed Forces Covenant reflects how policy-makers were forced to accommodate the 'legal component', emulating those senior Army officers involved in developing the Moral Component in the late 1990s. The military solution was the invention of the Military Covenant.

Conclusion

Writing after the Second Afghan War, General Sir Garnet Wolseley stated: 'The only rewards that are justly our [soldiers'] due are the gratitude of our country and the praise of our superiors'.[1] He cannot be accused of encouraging unrealistic expectations among the soldiers who had battled in Helmand during the 1880s. Gratitude and praise are a long way from the 'fair treatment' of the Armed Forces Covenant and the 'commensurate terms and conditions of service' that the Military Covenant says soldiers should receive. Written for the New Model Army, *The Soldier's Catechism* of 1644 states that soldiers should be 'well maintained with sufficient allowance'.[2] It adds:

> If they have received any hurt or loss by the warres [sic] ought to be liberally provided for and comfortably maintained all their days by them that sent them forth.[3]

From 2006, the Covenant became established in civilian society because soldiers serving in Iraq and Helmand received neither the material support demanded by the Catechism nor the gratitude that Wolseley considered just.

In five years, the Military Covenant went from being a paragraph in Army doctrine to statutory recognition. This study has sought to understand the institutional dynamics behind its development, the process of the concept's migration to the civilian sphere and the Covenant's impact on the civil–military relationship. The Covenant is a new development in the civil–military relations field, and one that, uniquely, in an area in which the United States has dominated since the 1950s, is British. Existing academic literature generally focuses on the post-migration Covenant, while taking for granted its establishment in the civilian sphere. Conversely, this study considers the Covenant's progress from the esoteric pages of Army doctrine to being incorporated in public policy is extraordinary: consequently its genesis and migration merited analysis.

The Military Covenant emerged as part of the development of the Army's Moral Component, a process that originally got underway in the mid-1990s but was finalised in 2000. This study sought to establish the catalyst for its development. Interviewees and analysts offer various explanations, ascribing factors both internal and external to the Army, including the cultural changes in

1 Wolseley, 1882, p. 2.

2 *The Soldier's Catechism*, p. 18.

3 Ibid., p. 28.

British society and their impact upon recruits (Dobbin, Wheeler); the need for the Army to assert its uniqueness prompted in part by the move towards 'jointery' (Roberts, Mileham); and, previously overlooked, the need for an ethical stock-take after a series of scandals that tarnished the Army's reputation (Ritchie, Andrews).[4] Forster suggests that the Covenant was the result of 'the threat of civilianization of the military' posed by the incorporation of the European Convention on Human Rights into British law.[5]

Bound up in the development of the Moral Component was an institutional exploration of the Army's 'right to be different', which evolved into the 'need' to be different during General Sir Roger Wheeler's tenure as CGS. However, it was the Army's *difference* from civilian society that was crucial, not least in the context of the development and codification of its institutional values and ethos. While a trend for articulating corporate values was also current in civilian society, the Army went much further, not least in setting out the values that soldiers were expected to evince.[6] Such prescription ran counter to the individualistic and non-judgmental mood of Blair-era civilian society. However, as Holmes subsequently suggested, the Army has a 'duty' to be different from civilian society, in particular to be 'value dominated'.[7] With the development of the Moral Component, the Army unwittingly attempted to fulfil Holmes' expectation of this duty.

Despite the regulations that govern every aspect of military life, soldiers, who offer 'unlimited liability', have no legal contract of employment. The emphasis the Army places on trust between the chain of command, along with its antipathy to contracts, accords with the institutional support for a covenant, an archaic form of trust-based agreement. This study suggests that the low-key influence of members of the Army Chaplain's Department was a factor in the development of the Moral Component, not least their interest in covenants as a means of ordering communities. *Soldiering*'s author, Sebastian Roberts, and the Covenant's principal advocate, General Dannatt, are men of profound religious faith. Without such faith, they might not have been familiar with – or unselfconsciously comfortable with – an ancient concept of little ostensible utility to a twenty-first-century organisation.

Paradoxically, before the Covenant's migration and promotion by civilian advocates, if they had heard of it, few soldiers gave the concept much thought. This study underlines that *Soldiering: The Military Covenant*, written for the military elite, was a background document that underpinned the core publication of the Moral Component, *Values and Standards of the British Army*, to which the Army accorded primacy. The Army had not attached too much institutional importance to *Soldiering* between 2000–06, reflected by its doctrinal demotion

4 From interviews, except Mileham, 2010, pp. 23–40.

5 Forster, 2012a, pp. 273–290.

6 While five corporations chosen at random from a Corporate Responsibility Index stress their corporate values, none details those expected of their employees (Anglian Water, BT, the Co-operative, EDF, Heineken), *Business in the Community*, 2012.

7 Holmes, 2006, p. 7.

in 2005. Conversely, *Values and Standards* was reissued to all soldiers in 2008 on the recommendation of the Aitken Report.[8] Many military interviewees, including *Soldiering*'s author, were surprised by the Covenant's success after 2006.

The Military Covenant came to national attention and reframed the civil–military relationship as a consequence of military operations in Iraq and Afghanistan. Following migration, the Covenant metamorphosed from a trilateral into a bilateral compact, which primarily conveyed the relationship between soldiers and government. Opinion was divided on the Covenant's condition, which CGS Dannatt, cautiously described as 'out of balance'. However, civilians, including academics, the media and politicians showed less restraint: 'broken almost beyond repair',[9] 'shattered by New Labour',[10] 'scandalously broken'.[11]

A civil–military 'gulf' between soldiers and civilian society is a theme of this study. Following the Iraq intervention, policy-makers and Forces personnel perceived the division between the two spheres had widened, a view strengthened by media reports of discrimination against, and abuse of, military personnel, most notoriously of wounded soldiers at Selly Oak hospital. In addition, policy proposals to bridge the gulf – suggested in the 2008 *National Recognition Report*[12] – as well as recurrent calls by policy-makers for legislation to outlaw discrimination against Forces personnel, reinforced the sense of an unprecedented civil–military divide. However, this study contends that the 'gulf' was a matter of perception. Media reports of anti-military sentiment were overblown, with many being implicitly anti-Moslem, an aspect overlooked by analysts. Public support for soldiers was latent: they were scapegoats for the controversial Iraq intervention and fell victim to the public's indifference to the mission in Afghanistan. Once the public was encouraged to decouple the men – and women – from the missions, this latent support began to manifest itself. The media and Help for Heroes, which sought and achieved celebrity endorsement, made our boys and our girls fashionable to back, reflected, for example, by the prominence University Hospitals Birmingham NHS Trust, home to Selly Oak, gave to its link with the Royal Centre of Defence Medicine after 2007.[13] Interviewees noticed the rapid shift in public attitude. The civil–military gulf was narrowed to the traditional norm, the gap. Dannatt stated that by the end of his tenure as CGS, the Covenant between soldiers and public had been 'brought back into balance',[14] a view with which this study concurs.

This study argues that the public's support for soldiers – that was extended to the Forces' community – was followed rather than led by policy-makers. With the Conservatives first unconditionally supporting the Government over Iraq, and then

8 MoD, 2008c.
9 Edmunds and Forster, 2007.
10 Fox, speech, 2010.
11 *The Independent*, 28 February 2010, p. 44.
12 MoD, 2008a.
13 UHB NHS Foundation Trust, 2006–07.
14 Dannatt, interview.

in the 2005 election incoherently trying to distance itself from that support, the Covenant finally gave the official Opposition the first opportunity to differentiate itself from Labour on defence policy in almost decade. However, in following an agenda set by General Dannatt and then by the Legion's *Honour the Covenant* campaign, MPs from all parties drew attention to their previously less than assiduous interest in 'people issues'. Despite the subsequent criticism, between the 1998 Strategic Defence Review and the 2003 Iraq intervention, Labour had made progress on welfare matters, neglected by the Major government keen to take the end-of-Cold War peace dividend. However, Iraq, inter-Departmental inertia and the chronic underfunding of defence meant that matters such as accommodation were imperfectly addressed. This study contends that the reframing of people issues in the context of the Military Covenant, forced the Brown government to re-focus on them, reflected by the 2008 Service Personnel Command Paper.[15]

With the bad luck that would dog his premiership, Gordon Brown came to office weeks before the launch of Help for Heroes and the *Honour the Covenant* campaign. On the wrong side of the Army, the Legion, military charities, coroners and the media when British Service personnel were involved in combat operations, the government was also its own worst enemy. Appointing Des Browne Defence Secretary and Scottish Secretary sent a message that defence was a part-time job. In addition, this study suggests that the Brown government appeared oblivious to the Covenant's inherent moral force. The Government's refusal to recognise the concept did it no favours, especially when the Covenant became, as Tipping described, a 'portmanteau' term and its scope was extended to include operational matters such as a reported shortage of helicopters. As the government defended itself and justified the conflicts in policy-jargonese, upbeat statistics and percentages, the public saw hearses carrying flag-draped coffins process through Wootton Bassett and could often read that the dead were as much victims of government parsimony as of the Taliban or Iraqi insurgents.

A paragraph of military doctrine based on the moral principle of reciprocity, the Military Covenant became a weapon with which the Army entered the political sector. In response to whether General Dannatt had used the Covenant in a 'tactful' way to convey the pressure the Army was under, *Soldiering*'s author suggests that the concept's use had been 'tactical'.[16] Historically, the long-standing fear in Britain that a standing army would lead to Caesarism and another civil war, meant that after the 1688 Settlement, as Howard suggests, 'Britain's tiny army remained the object of quite ludicrous suspicion'.[17] Today, some media commentary seems similarly influenced by the 1650s Rule of the Major Generals, with an over-excitable tendency to assume that any contemporary military incursion into politics signals that the 'supplantment', described by Finer, is imminent.[18]

15 MOD, 2008b.

16 Roberts, interview.

17 Howard, 1957, p. 14.

18 Finer (introduction Stanley), 1992.

If citizens in Britain today are troubled by the dichotomy between their liberty and their security identified by Burke, their attention should focus on civilian agents rather than the military. Strachan suggests that historically, far from being apolitical, the Army was always an integral part of the state and policy-making.[19] The advent of social media has not only created a more participatory news culture, as Loader and Mercea observe,[20] but is integrating the military more deeply in the civilian sphere. With social media, all soldiers now have the potential to make a strategically or politically significant statement. The image that, for many, is associated with Iraq is not a *Guernica*-style painting, but a photograph of prisoners in Abu Ghraib. The MoD media blackout on Helmand in 2006 was by-passed with footage from soldiers' helmet-cams, while *Sun* readers could see for themselves the state of soldiers' accommodation.[21] Inevitably, soldiers will be politically engaged; concerns can be aired, monitored and acted upon via websites such as the Army Rumour Service. *Values and Standards* states that 'our responsibility to the Crown through Parliament, means that the Army must be apolitical and be seen to be apolitical'.[22] In the social media era, expecting soldiers to stay publically aloof from politics is unrealistic.

CGS Dannatt's entry in the political sector reflected and exacerbated the existing disconnect between the Government and the Armed Forces. This aspect of the gulf in the civil–military relationship demands more analysis: with soldiers involved in combat operations, it cannot be helpful if senior commanders and the government are also at war. Dannatt's tenure as CGS deserves greater appraisal: not only did he promote the cause of soldiers' welfare and was popular throughout the chain of command for doing so, but as this study confirms, one Labour defence minister believes he should have been sacked following his *Daily Mail* interview, as did some media commentators.[23] Suspicion that he was a 'Tory stooge' appeared to be confirmed by his appointment as a Cameron advisor, which, along with the attack on the Government by five former Chiefs in the House of Lords in November 2007, suggested the Armed Forces had lost their political neutrality. In the United States, civil–military relations analysis was galvanised by the 2006 'Revolt of the Generals' against Secretary Rumsfeld and the 2010 sacking of Stanley McChrystal, whose insubordination was minor compared with that shown by Dannatt.

The Covenant became a partisan concept, adopted and used for party political advantage. Many military interviewees spoke of politicians 'hijacking' it. While some might describe the Opposition parties' call to support the Forces' community as altruistic concern, others might see it as self-serving opportunism.

19 Strachan, 1997.

20 Loader and Mercea, 2011, pp. 757–769.

21 Newton Dunn, *The Sun*, 26 October 2007, p. 1.

22 ADP, 2000b, paragraph 29. Army Doctrine Publications. Ministry of Defence. © Crown copyright 2014.

23 Kevan Jones MP, interview; Matthew Parris, Max Hastings, Simon Jenkins.

After migration, the Covenant became a 'dog-whistle' term. Politicians could use it, silently conjuring up military sacrifice, while avoiding accusations of exploitation, thereby retaining the moral high ground. This was especially apparent after the *Honour the Covenant* campaign, as all parties scrambled to show concern for the Forces. However, the parties were united that a 'gulf' between the Armed Forces and society existed: policy documents give the impression that policy-makers had somehow failed to notice that the public was indeed honouring the Covenant and demonstrating their appreciation for soldiers, if not for the operations in which policy-makers had involved them.[24] Tut-tutting over the public's imaginary backsliding, policy-makers forgot that only they could ensure soldiers received the 'commensurate terms and conditions of service' which the Military Covenant demanded.

This study opened by seeking to place in a historical context the Covenant's fundamental principle – support from the Army and nation (both the public and policy-makers) in exchange for soldiers' service and sacrifice. While the Covenant codifies that a reciprocal bond of 'identity, loyalty and responsibility' has sustained the Army 'throughout its history', this claim has also been advanced by some of the concept's civilian champions, including the Royal British Legion. Tommy Atkins, alternately shunned and revered by late Victorian society, has become a stock character, who, for many, still typifies public attitudes towards the military. However, Kipling's assertions are cited rather than tested. Today, the ubiquity of war memorials to the ordinary soldier – and sailor and airman – rather than to their commanders, along with the annual commemoration of the fallen and the Poppy appeal, implies that the nation recognises its debt to those who served in the twentieth century. However, such memorialising can obscure the post-1918 reality for some veterans who were promised 'a land fit for heroes', but whose pensions remained discretionary rather than a right. Disabled veterans continued to serve the nation by putting on a uniform of good cheer at institutions such as the Star and Garter home, rather than supporting political extremism like their former adversaries in Europe.[25] After 1945, many civilians showed a lack of enthusiasm towards returning Service personnel,[26] while, subsequently, National Service often provided comic material for light entertainment rather than inspiring reverence for the Armed Forces. Once peace breaks out, despite memorials to the dead, the nation's moral debt to dependents and surviving veterans is not always discharged. This study concludes that, although soldiers themselves received greater public approbation from the late nineteenth century when they came to be perceived as agents of benign imperialism, historically, the nation's sense of moral or material obligation towards them has always been imperfectly fulfilled.

---

24 E.g. Each and every one of us has a responsibility to do more to support the men and women of our Armed Forces. SDSR, 2010, p. 6.

25 Cohen, 2001.

26 Allport, 2009, pp. 50–80.

One challenge to any assessment of the Covenant's historical claims is the scale of the period under review: 350 years is a great deal of ground to cover and treating the Army as a holistic entity might not find favour among too many military historians. Today's Army is very different from that of the early Cold War conscription era, let alone from the remnants of the New Model from which some regiments can trace their origins. French or Keegan might argue that it is Britain's regimental system that has better provided soldiers with their military identity and, historically, has 'looked after its own' – whether veterans or their dependents – in case of disability or death.[27] The Royal Engineers Officers Widows Society, for example, was established in 1785, specifically to provide for the dependents of Sapper officers.[28] The much-cited 1593 Act for Relief of Soldiers, does not withstand scrutiny as being indicative of a national sense of altruistic obligation to those who had served:[29] it was a public order measure that foisted responsibility for veterans' welfare onto local parishes. The Covenant's historical claims are overblown.

The formalisation of doctrine by the Army in 1989 was a departure for an institution that prided itself on the ability of its senior officers to improvise, or less kindly, muddle through. As Dannatt observed, during the 1980s the Army underwent an explosion of thinking and doctrinal development; by 2000, it had 'a full hand of doctrines and concepts'.[30] Some interviewees questioned the perceived absence of doctrine, pointing out that one definition of doctrine is 'that which is taught'. The impact of the formalisation of doctrine after 1989 on British military performance could be subject of future academic analysis.

Although the Moral Component can be seen as an attempt to build a 'moral community', the Army's civilian critics, particularly some Labour MPs, saw it as anything but, not least because of the apparent institutional antipathy to equal opportunities. In highlighting the civilian encroachment on the Army's professional space, Forster suggests that senior commanders should have done more to defend it by not wasting their metaphorical bullets on issues arising from the Human Rights Act.[31] The Army – rather like its Commander-in-Chief at the time of the death of Diana, Princess of Wales – found itself culturally out of step with Blair's Britain. With hindsight, it seems curious that it did not take the opportunity to align itself better with the pre-9/11 *zeitgeist* and promote itself as an upholder of international humanitarian norms, whether in Bosnia, Kosovo or Sierra Leone.

This study establishes that media was crucial to the Covenant's migration. An analysis of newspaper coverage reflects how soldiers' welfare came to be seen through the Covenant prism, particularly after August 2007 when the Legion

27 French, 2004.

28 Royal Engineers Officers' Widows Society, army website.

29 Those citing the 1593 Act include the Legion, the Conservative Party, the BBC, and Haddon-Cave.

30 Dannatt, speech, 21 September 2007.

31 Forster, 2006, pp. 1043–1057.

announced the forthcoming launch of its *Honour the Covenant* campaign. The Covenant concept was generally confined to what could be described as the civilian 'officer class' of policy-makers and opinion-formers in the broadsheet press. Its name capturing exactly what it wished to do, Help for Heroes resonated immediately among the wider public, not least because of support from celebrities, sports stars and *The Sun*. Military charities have traditionally been a conduit for the public to honour their obligations to the Forces reflected by initiatives such as the Crimean War-era Royal Patriotic Fund. The media encouraged expressions of public appreciation for soldiers, who in colloquial terms, went from zeroes to heroes after 2007. Whether the media form or reflect opinion is always debatable, however they are rarely out of tune with the public. Dannatt's successful deployment of the Covenant led to the concept and the people issues associated with it – including housing, pay, medical care – becoming the dominant issues in connection with defence, for example superseding far-reaching policy decisions such as the commission of two new aircraft carriers.[32] The informal ceremony of repatriation at Wootton Bassett became central to promoting national awareness of soldiers' sacrifice. Decried as mawkish,[33] Wootton Bassett demands further study, not least for its effect on public opinion. Together with the issue of military victimhood – i.e. the misperception by many in the civilian sphere that a disproportionate number of Service personnel end up in prison, sleeping rough, suffering from PTSD or committing suicide – the Wootton Basset phenomenon could sap support among an increasingly risk-averse British public for military operations in the future.

Soldiers' service and sacrifice has been tangible since 2000, with, to date, more than 600 dead and thousands wounded in Iraq and Afghanistan. This study suggests that the focus on the Covenant after 2006 allowed questions concerning recent British military performance to be deferred. Testimony by senior commanders and politicians to the Chilcot Inquiry underlines not just the ad hoc nature of Operation Telic, but the folly of Herrick. The initial rapid deployment of 46,000 personnel to Iraq in early 2003 was a success, especially in the light of the eleventh hour uncertainty about whether ships were going to 'turn left or right' in the Mediterranean.[34] Analysts seeking to explain the Army's willingness to be part of the expanded NATO mission in Afghanistan in 2006 have suggested various institutionally self-serving reasons, including the Northern Ireland drawdown and the need to redeem its reputation following a less than glorious operation in Iraq.[35] If senior commanders placed the Army's needs ahead of the nation's they broke the Covenant. Britain's Ambassador to Afghanistan, who was taken to task for the pessimism of his dispatches to London, said that all he heard was 'happy talk' from many senior officers and a familiar refrain: 'The *leitmotiv* of a long war: we are

32 Sands, 2006, p. 12.

33 (General Sir Robert Fry) Coghlan, *The Times*, 13 November 2010.

34 Boyce – Chilcot, p. 50.

35 E.g. Ledwidge; Haynes et al., *The Times*; Cowper-Coles.

making progress, but challenges remain'.[36] Senior commanders were as adept at spin as any of the Labour ministers they were serving, which raises the question of the quality of their appraisals of progress in either Iraq or Afghanistan. Success is far harder than victory to determine – especially when the objectives keep changing.

The opacity surrounding the justification for the intervention in Iraq in 2003 as well as the incoherence of Herrick's objectives is mirrored by the lack of clarity surrounding the decision-making process about the double deployment, for which neither politicians nor senior commanders now wish to take the credit. The Army's commitment to upholding its Covenant with soldiers is called into question with the admission by former CDS General Lord Walker that some soldiers' lives would be made 'even more intolerable' with the continuing breach of harmony guidelines that the double deployment would involve.[37] In addition, the most emotive perceived breach of the Covenant concerned the shortfall in equipment, notably helicopters, depicted by government critics as concerning, primarily, force protection rather than military utility. Blame can be attributed to the Labour government for not properly funding defence following the 1998 Strategic Defence Review: however, decisions about procurement also rest with Armed Forces themselves. The Future Rapid Effects System (FRES), fast jets and carriers, which all had their Service advocates, were consistently preferred over less glamorous items such as helicopters. The Commander of Herrick 4 observes that 'insufficient lessons were learnt from Iraq',[38] two of which could be vulnerability of Snatch Land Rovers and the utility of UAVs, iterated by the Defence Select Committee in 2002.[39] While senior commanders often give the impression that the Forces are constitutionally helpless, they have a duty to the nation not to succumb to the 'conspiracy of optimism' about military success that has done defence procurement such disservice.

The Military Covenant is often intertwined with the legal component that increasingly confronts both the Army as an institution and individual soldiers. Indeed, Forster suggests that the Covenant's codification was, in part, a response to the growing incursion of civilian law into the Army's professional space.[40] However, such incursions have continued since the 2000 codification, with the concepts of combat immunity and Crown immunity being challenged, reflected by, for example, the uncertainty surrounding the extent of the MoD's legal duty of care in protecting soldiers on operations.[41] The legal component frequently tests the strength of the Covenant's bonds of loyalty and responsibility. The failure of a series of prosecutions against soldiers – Trooper Williams, seven members of 3 Para and the court martial of members of the Queens Lancashire Regiment – gave

36 Cowper-Coles, 2011.

37 Walker, 19 Sept 2005 (Chilcot – written evidence); Walker, Chilcot pp. 56–57.

38 Butler, interview.

39 HC 91–1 2002–03, paragraph 111.

40 Forster, 2006, pp. 1043–1057.

41 A perimeter fence: Supreme Court, July 2010.

credence to claims that an anti-military government was pursuing a politically correct agenda against soldiers, despite committing them to a war of questionable legality that, according to some analysts, also failed to meet the 'just war' criteria.[42] As Tootal suggests: 'Combat is a pretty extreme place to be. Especially when soldiers are being zipped into body bags'.[43] With opinion polls consistently underlining the public's goodwill towards the Armed Forces, it can be assumed that the public would agree with him. However, along with the observation made in 2005 by former CDS, Lord Boyce, that members of the Armed Forces were under 'legal siege', were complaints from soldiers being prosecuted that they were 'hung out to dry' by the Army, which hardly suggests the Covenantal bonds between the chain of command were robust in this instance.[44] Conversely, like Deepcut, the Al-Sweady and Baha Mousa inquiries raise questions about whether individual soldiers were honouring the Covenant. Many interviews for this study were conducted while the Coalition government attempted to give it statutory recognition. Military interviewees were concerned that a statutory Covenant could be justiciable, which could undermine combat effectiveness and put the Army in the dock.

The statutory recognition given to the Military Covenant in the 2011 Armed Forces bill is just one aspect of its entrenchment in public policy. The judiciary has acknowledged the Covenant independently and made concrete the concept's principle of support in exchange for military service and sacrifice. In the most far-reaching examples, the Covenant was integral to a 2008 immigration judgment involving Gurkha veterans and the 2009 Haddon-Cave Inquiry in which the duty of care civilian contractors owe to Service personnel was examined. While the Conservative Opposition and the Coalition government trumpeted their intention to 'enshrine the Military Covenant in law', they seemed unaware of these developments. The Covenant's ever-deeper entrenchment in policy is, paradoxically, treated with wariness by military interviewees, alert to that danger that, should the Forces' community be treated as a privileged caste, the current goodwill towards them might evaporate. The principle of 'no disadvantage' was meant to quell any notion that soldiers were 'citizens-plus'.[45]

The Armed Forces Covenant is an anodyne, lawyer-proof imitation of the Military Covenant. In government, the Conservatives found themselves boxed in by their own rhetoric while in Opposition. The penny dropped, slowly, that any agreement that offered Forces' personnel 'commensurate terms and conditions of service' in exchange for their unlimited liability would be tested in court. Attempting to retreat from its undertaking to give the Covenant statutory recognition, the government then appeared to buckle under moral pressure from the Legion and, somewhat ironically, from the now defunct *News of the World*, which cited the

42 E.g. Quinlan, 2003; Fisher, 2011.

43 Tootal, interview.

44 Boyce, 14 May 2005, HL c1236; including Tim Collins, Jorge Mendonca.

45 Officer G, interview.

'legal' Covenant as one of its final achievements. The Armed Forces Covenant and the measures that accompany it are not much of an advance on initiatives within Labour's 2008 Command Paper. Many interviewees are sceptical about the concept, believing it to be little more than Coalition camouflage to obscure defence cuts and redundancies among Forces' personnel, about which the Army Families Federation received 2000 emails in five days.[46] While wanting to give the government the benefit of the doubt, Prime Minister Cameron's 'you do the fighting and I'll do the talking' angered many military interviewees,[47] and is hardly reflective of the 'respect and support' codified in the Coalition government's Armed Forces Covenant of 2011.

The Military Covenant's migration and subsequent acknowledgement as formal component of public policy has coincided with the questioning of the market orthodoxy that has prevailed in Britain and the United States in recent decades. Disenchantment with the neo-liberal agenda has been exacerbated by the financial crisis that began in 2007, the accompanying bank bailouts and the moral hazard associated with them. Concurrent with market failure was ethical failure, represented by the Parliamentary expenses and the phone-hacking scandals. Both the Conservative Party's Big Society and 'Blue Labour' initiatives call for the renewal of civil society and seek to foster civic virtue, trust and community. Alasdair Macintyre has influenced Blue Labour luminary Maurice Glasman, as well as members of the Army Chaplaincy, particularly Torrance, who were involved in developing the Moral Component. In establishing itself as a 'moral community' by emphasising its values and standards, the Army differentiated itself from civilian society. However, having previously been castigated with being out of step with Blair's Britain, reflected by, for example, its attitudes towards minorities, before too long the Army could be perceived as an ethical exemplar, particularly as demands grow for civilian society to renew its own moral component.

The Military Covenant's journey from Army doctrine to statutory recognition has coincided with the toughest war-fighting British soldiers have experienced since Korea. It came to symbolise the privations suffered by those soldiers – underpaid, poorly equipped, under-resourced, overstretched – but it was also the means to right some of the wrongs. Since 2007, the British public has shown it wanted to honour the Covenant, not least at Wootton Bassett. A timeless statement, *Soldiering: The Military Covenant* anticipated soldiers' service and sacrifice in Iraq and Afghanistan. With its associations with the sacred, the Covenant is a reminder that for some soldiers, what they do is not just a profession, but a vocation. Many soldiers only became aware of the Military Covenant once the Army had lost ownership of it. As one said: 'When we had it, we didn't value it'.[48] However, the Military Covenant still best represents the obligations owed to soldiers who are prepared to serve the nation and make the ultimate sacrifice.

46 McCarthy, interview.

47 Including Officer L, interview.

48 Officer L, interview.

Bibliography

Abbott Parry, Edward and Codrington, Alfred 1918. *War Pensions: Past and Present.* London: Nisbet and Co.

Act for the Necessary Relief of Soldiers and Mariners, 1592. Parliamentary Archives, Houses of Parliament HL/PO/PU/1/1592/35Eliz1n4.

Act for the Necessary Relief of Soldiers and Mariners, made in the 35th Year of the Queen's Majesty's Reign that now is, 1597. Parliamentary Archives, Houses of Parliament HL/PO/PU/1/1597/39Eliz1n21.

Act for the Necessary Relief of Soldiers and Mariners, 1601. Parliamentary Archives, Houses of Parliament HL/PO/PU/1/1601/43Eliz1n3.

Adams, Paul 2007. Is Britain Failing its Armed Forces? BBC News (online), 13 October. Available at: www.bbc.co.uk/ethics/war/overview/covenant.shtml (accessed 6 June 2010).

Adams, Paul and Simpson, Aislinn 2007. Parade Snub for Britain's Returning Soldiers, *Daily Telegraph* (online), 26 September. Available at: http://www.telegraph.co.uk/news/uknews/1564191/Parade-snub-for-Britains-returning-soldiers.html (accessed 14 June 2010).

Ainsworth, Rt Hon Bob, MP 2009. *Courage and Commitment: Doing the Job in Afghanistan, Supporting Personnel at Home.* Speech to RUSI, London, 17 December.

Air Staff (MoD) 1993. *Air Power Doctrine*, AP3000, 2nd edition.

Aitken, Robert 2008. *An Investigation into Cases of Deliberate Abuse and Unlawful Killing in Iraq in 2003 and 2004.* Ministry of Defence. Available at: http://www.mod.uk/DefenceInternet/DefenceNews/DefencePolicyAndBusiness/ModReleasesReportIntoAbuseOfIraqiCivilians.htm (accessed 17 November 2010).

Allphen Moore, John Jr. and Pubantz, Jerry 2002. *Encyclopedia of the United Nations.* New York: Infobase Publishing.

Anderson, Benedict 2006. *Imagined Communities* (revised edition). London: Verso.

Anderson, Olive 1971. The Growth of Christian Militarism in Mid-Victorian Britain, *English Historical Review*, 86:338, Jan, pp. 46–72.

Andrews, Major General Stephen, CBE 2011. Interview with author, London.

Aristotle 1925 and 1980. *The Nicomachean Ethics* (trans. David Ross, revised J.L. Ackrill and J.O. Urmson). Oxford: Oxford University Press.

Armed Forces Select Committee 2000–01. *Special Report*, HC 154-I.

—— 2005–06. *Special Report*, HC 828.

—— 2010–12. *Official Report, Armed Forces Public Bill Committee*, HC 779.

Army Doctrine Publication 1989. *Design for Military Operations: The British Military Doctrine*, AC 71451 (Liddell Hart Centre for Military Archives

(LHCMA), British Army Field Manuals and Doctrine Publications Collection. 1963–2004 GB 0099).

—— 1996. *Design for Military Operations: The British Military Doctrine*, AC 71415, 2nd edition (LHCMA).

—— 2000a. Volume 5, *Soldiering: The Military Covenant*, AC 71692 (LHCMA).

—— 2000b. *Values and Standards of the British Army* (Commanders' Edition), AC 63813 (LHCMA).

—— 2000c, *Values and Standards of the British Army* (Soldiers' Edition). AC 63812 (LCHMA).

—— 2005. *Land Operations*, AC 71819.

—— 2008. *Values and Standards of the British Army.* Available at: http://www.army.mod.uk/documents/general/v_s_of_the_british_army.pdf (accessed 17 November 2010).

—— 2010. *Operations*, AC 71632. Available at: https://www.gov.uk/government/uploads/system/uploads/attachment_data/file/33695/ADPOperationsDec10.pdf (accessed 8 March 2012).

Aylwin Foster, Nigel 2005. Changing the Army for Counter-Insurgency Operations, *Military Review*, 85:6, Nov/Dec, pp. 2–15.

Babington, Anthony 1990. *Military Intervention in Britain: From the Gordon Riots to the Gibraltar Incident.* London: Routledge.

Bacevich, Andrew J. 2005. *The New American Militarism: How Americans Are Seduced By War.* Oxford: Oxford University Press.

Baier, Annette 1986. Trust and its Varieties, *Ethics*, 96, pp. 231–260.

Bailey, Charlotte 2008. MoD Poll: Public's Support for Armed Forces at Record Levels. *Daily Telegraph*, 1 Nov. Available at: http://www.telegraph.co.uk/news/newstopics/onthefrontline/3333357/MoD-poll-Publics-support-for-Armed-Forces-at-record-levels.html (accessed 2 February 2011).

Bailey, Tom. On Trust and Philosophy, *Open University*, BBC. Available at: www.open2net/trust/on_trust/on_trust1.

Baldwin, Tom and Evans, Michael 2008. US Accuses Britain over Military Failings in Afghanistan, *The Times*, 16 December, p. 3.

Barakso, Mary Ann and Schaffner, Brian F. 2006. Winning Coverage: News Media Portrayals of the Women's Movement 1969–2004, *Harvard International Journal of Press/Politics*, 11:22, pp. 22–44.

Batty, David 2009. Commander Issued Helicopter Warning Weeks Before Afghanistan Death, *The Guardian*, 31 October. Available at: http://www.guardian.co.uk/uk/2009/oct/31/rupert-thorneloe-helicopter-shortage-afghanistan (accessed 2 November 2009).

Baugh, Daniel A. 1998. Great Britain's Blue-Water Policy, *International History Review*, vol. X, pp. 33–58.

Baylis, John (ed.) 1977. *British Defence Policy in a Changing World.* London: Croom Helm.

BBC News 2006a. Deepcut Families Vow to Fight On, 29 June, BBC (online). Available at: http://news.bbc.co.uk/1/hi/uk/4857990.stm (accessed 28 March 10).

—— 2006b. TV's Snow Rejects Poppy Fascism, BBC (online), 10 November. Available at: http://news.bbc.co.uk/1/hi/uk/6134906.stm (accessed 30 July 2011).
—— 2008a. David Cameron: Military Covenant Broken, 4 August. Available at: http://news.bbc.co.uk/1/hi/uk_politics/7276924.stm (accessed 23 January 2011).
—— 2008b. Forces Tribute with Fly-Past, 12 October. Available at: http://news.bbc.co.uk/1/hi/uk/7666109.stm (accessed 2 August 2011).
—— 2009a. Thousands Pay Respect to Soldiers, 14 July. Available at: http://news.bbc.co.uk/1/hi/uk/8150938.stm (accessed 14 July 2009).
—— 2009b. Wootton Bassett Sheds its Tears for Soldiers, 10 November. Available at: http://news.bbc.co.uk/1/hi/uk/8353290.stm (accessed 10 November 2009).
BBC Politics Show 2009. UK 'not convinced' by Afghan goal, 8 November. Available at: http://news.bbc.co.uk/1/hi/uk/8348942.stm (accessed 15 January 2011).
BBC/ICM 2007. *Iraq Survey* March. Available at: http://news.bbc.co.uk/1/hi/6467147.stm (accessed 15 January 2011).
BBC2 2009. *Newsnight*, Haddon-Cave Report, 28 October 2009 (viewed 28 October 2009).
Beale, H.G., Bishop, W.D. and Furmston, M.P. 2008. *Contract: Cases and Materials*, 5th edition. Oxford: Oxford University Press.
Beevor, Antony 1990. *Inside the British Army*. London: Chatto and Windus.
Bell, Christopher M. 2005. The Royal Navy and the Lessons of the Invergordon Mutiny, *War in History*, 12, pp. 75–92.
Bell, Martin 2007. This is a Nation Debt. It Must Be Repaid, *Mail on Sunday* (online), 25 March. Available at: http://www.dailymail.co.uk/news/article-444435/Martin-Bell-This-national-debt-It-paid.html (accessed 22 March 2011).
Bett, Michael 1995. *Independent Review of the Armed Forces' Manpower, Career and Renumeration Structures*, HMSO.
Betz, David and Cormack, Anthony 2009. Iraq, Afghanistan and British Strategy, *Foreign Policy Research Institute*, Spring.
Binns, Major General Graham CBE DSO MC 2010. Evidence to Chilcot Inquiry, 15 January.
Birchall, Jonathan 2009. Robot Hamsters Hottest Toys this Christmas, *Financial Times*, 8 November, p. 1.
Bishop, Patrick 2007. *3 Para*. London: Harper Perennial.
Blair, Rt Hon Tony, MP 1999. *Doctrine of the International Community*, Economic Club of Chicago, Chicago, 22 April.
—— 2007. *Our Nation's Future – Defence*, HMS *Albion*, Portsmouth, 12 January 2007.
—— 2010. *A Journey*. London: Hutchinson.
—— 2010. Evidence to Chilcot Inquiry, 21 January.
—— 2011. Evidence to Chilcot Inquiry, 29 January.
Blake, Sir Nicholas, QC 2006. *A Review of the Circumstances Surrounding the Deaths of Four Soldiers at the Princess Royal Barracks, Deepcut, Between 1995 and 2002*.

Blau, Peter M. 1986. *Exchange and Power in Social Life.* New Brunswick: Transaction.

Bond, Brian 2002. *The Unquiet Western Front: Britain's Role in Literature and History.* Cambridge: Cambridge University Press.

Bonehill, John and Quilley, Geoff (eds) 2005. *Conflicting Visions: War and Visual Culture in Britain and France c1700–1830.* Aldershot: Ashgate.

Booker, Christopher 2011. War Hero Pex Wins the Right to Stay, *Sunday Telegraph*, 28 May 2011.

Boone, Jon, Steel, Jonathan and Norton-Taylor, Richard 2010. WikiLeaks Cables Expose Afghan Contempt for British Military, *The Guardian*, 2 February. Available at: http://www.guardian.co.uk/uk/2010/dec/02/wikileaks-cables-afghan-british-military (accessed 2 February 2010).

Bowen, H.V. 1998. *War and British Society, 1688–1815.* Cambridge: Cambridge University Press.

Boyce, Rt Hon Admiral the Lord, KG GCB OBE DL 2009. Evidence to Chilcot Inquiry, 3 December.

—— 2010. Evidence to Chilcot Inquiry, 11 January.

Bray, Mike 2011. Interview with author, London.

Brereton, J.M. 1886. *The British Soldier: A Social History from 1661 to the Present Day.* London: Bodley Head.

Brewer, John 1989. *The Sinews of Power.* London: Unwin Hyman.

Briner, Rob and Conway, Neil 2005. *Understanding Psychological Contracts at Work: A Critical Evaluation of Theory and Research.* Oxford: Oxford University Press.

Brown, Anthony 1971. *Red for Remembrance: The British Legion 1921–71.* London: Heinemann.

Brown, Lt-Gen Chris 2011. *Operation Telic Lessons Compendium*, April 2011. Available at: http://www.mod.uk/NR/rdonlyres/F0282A90–99E5–415E-B3BC-97EAD7D7873A/0/operation_telic_lessons_compendium.pdf (accessed 21 October 2011).

Brown, Rt Hon Gordon, MP 2010. Evidence to Chilcot Inquiry, 5 March.

Browne of Ladyton, Lord (Des) 2010. Evidence to Chilcot Inquiry, 25 January.

Burroughs, Peter 1985. Crime and Punishment in the British Army 1815–1870, *English Historical Review*, 100:396, pp. 545–571.

Business in the Community 2012. *Corporate Responsibility Index.*

Butler, Brigadier (Retired) Ed, CBE DSO 2011. Telephone interview with author.

Cabinet Office 2010a. *The Coalition: Our Programme for Government*, 401238/0510.

—— 2010b. *Strategic Defence and Security Review: Securing Britain in an Age of Uncertainty.*

—— 2010c. *A Strong Britain in an Age of Uncertainty: The National Security Strategy.*

Caforio, Guiseppe 2006. *Handbook of the Sociology of the Military.* New York: Springer.

Callwell, C.E. 1899. *Small Wars: Their Principles and Practice.* London: Harrison and Sons on behalf of HMSO.

Cameron, Rt Hon David, MP 2009. *The Big Society.* Hugo Young Lecture, London, 10 November. Available at: http://www.conservatives.com/News/Speeches/2009/11/David_Cameron_The_Big_Society.aspx (accessed 1 December 2011).

—— 2010. HMS *Ark Royal*, Nova Scotia, 24 June. Available at: http://www.number10.gov.uk/news/pms-speech-in-halifax-nova-scotia/ (accessed 26 June 2010).

—— 2010. *Now We Must Do Our Duty By Servicemen and Women.* Sunday Telegraph, 19 June 2010, p. 17.

Campbell, David (ed.) 2001. *The Relational Theory of Contract: Selected Works of Ian Macneil.* London: Sweet and Maxwell.

Campbell, Rt Hon Sir Menzies, MP QC 2008. *No Choice But Change: The Military Covenant in its Strategic Context.*

—— 2012. Interview with author, London.

Chandler, David (ed.) and Beckett, Ian (associate ed.) 1994. *The Oxford Illustrated History of the British Army.* Oxford: Oxford University Press.

Chartered Institute of Personnel Development 2009. *The Psychological Contract.* Available at: www.cipd.co.uk/subjects/empreltns/psycntrct/psycontr.htm (accessed 12 November 2009).

Clarke, Michael 2007. Strategy and Fortune, *RUSI Journal*, 152:5, Oct, pp. 6–12.

—— 2011. *The Afghan Papers.* London: RUSI.

Codner, Michael and Clarke, Michael 2011. *A Question of Security: The British Defence Review in an Age of Austerity.* London: IB Taurus.

Coghlan, Tom 2010. General Fears 'Mawkish' View of Military; Veterans Unhappy with Celebrity Subversion of Armistice Day, *The Times*, 13 November, p. 3.

Cohen, Deborah 2001. *The War Come Home: Disabled Veterans in Britain and Germany 1914–1939.* Berkeley and Los Angeles: University of California Press.

Coletta, Damon 2007. Courage in the Service of Virtue: The Case of General Shinseki's Testimony to Congress, *Armed Forces and Society*, October.

Colley, Linda 1984. The Apotheosis 3of George III: Loyalty, Royalty and the British Nation, 1760–1820, *Past and Present*, 102, pp. 94–129.

—— 1992. *Britons.* London: BCA by arrangement with Yale University Press.

Connelly, Mark 2002. *The Great War, Memory and Ritual.* Woodbridge: Royal Historical Society Boydell Press.

Conservative Party 2008. *Restoring the Covenant: The Military Covenant Commission's Report to the Leader of the Conservative Party.*

—— 2010a. *Armed Forces Manifesto: A New Covenant For Our Armed Forces and their Families.*

—— 2010b. National Citizen Service. Available at: http://www.conservatives.com/News/News_stories/2010/04/~/media/Files/Downloadable%20Files/NCSpolicypaper.ashx. (accessed 14 February 2011).

Cook, Rt Hon Robin, MP 2003. *Why I had to leave the Cabinet*, The Guardian, 18 March, p. 26.

Cooper, Lt Gen John, CB, DSO MBE 2009. Evidence to Chilcot Inquiry, 12 December.

Cornish, Paul 2010. *Strategy in Austerity: The Security and Defence of the United Kingdom.* Chatham House Report, October. Available at: http://www.chathamhouse.org/sites/default/files/r1010_cornish.pdf (accessed 30 October 2010).

Cornish, Paul and Dorman, Andrew 2009. Blair's Wars and Brown's Budgets: From Strategic Defence Review to Strategic Decay in Less Than a Decade, *International Affairs*, 85:2, pp. 247–261.

Cowper-Coles, Sir Sherard, KCMG LVO 2010. Supplementary Written Evidence, Foreign Affairs Select Committee, 2010–12, *The UK's Foreign Policy Approach to Afghanistan and Pakistan* HC 514.

—— 2011. *Cables from Kabul.* London: Harper Collins.

Coyle-Shapiro, J. and Kessler, Ian 2000. Consequences of the Psychological Contract for the Employment Relationship: A Large-Scale Survey, *Journal of Management Studies*, 37:7, pp. 903–930.

Croft, Stuart, Dorman, Andrew, Rees, Wyn and Uttley, Matthew 2001. *Britain and Defence 1945–2000: A Policy Re-evaluation.* Harlow: Pearson Education.

Cross, Tim, Major General CBE 2009. Evidence to Chilcot Inquiry, 7 December.

—— 2012. Interview with author, London.

Cruikshank, C.G. 1946. *Elizabeth's Army.* Oxford: Oxford University Press.

Currie, Major General Peter (Retired) CB 2011. Interview with author, London.

Curtis, Nicola 2009. Mayor and British Legion Hit Out at Wootton Bassett Question Time, *Wiltshire Gazette and Herald* (online), 9 December. Available at: http://www.gazetteandherald.co.uk/news/headlines/4782796.Mayor_and_British_Legion_hit_out_at_Wootton_Bassett_Question_Time/ (accessed 18 July 2011).

Daddow, Oliver and Gaskarth, Jamie (eds) 2010. *British Foreign Policy: The New Labour Years.* Basingstoke: Palgrave Macmillan.

Daily Mail Comment 2007. Our Boys Betrayed, *Daily Mail*, 14 September, p. 14.

Daily Mail Comment 2007. Shameful Failure to Care for our Forces, *Daily Mail*, 12 March, p. 14.

Daily Mail Comment 2010.Wounded Heroes and a Question of Honour, *Daily Mail* (online), 6 September. Available at: http://www.dailymail.co.uk/debate/article-1309381/DAILY-MAIL-COMMENT-Wounded-heroes-question-honour.html (accessed 1 July 2011).

Daily Mail Reporter 2006. Harrods Bans Soldiers on Poppy Day, *Daily Mail* (online), 18 November. Available at: http://www.dailymail.co.uk/news/article-417249/Harrods-bans-soldiers-Poppy-Day.html (accessed 11 November 2010).

Daily Mail Reporter 2007. Asian Petrol Station Worker 'Refused to Serve Afghanistan War Veteran Because He Was in Uniform', *Daily Mail* (online), 24 October. Available at: http://www.dailymail.co.uk/news/article-489407/Asian-petrol-station-worker-refused-serve-Afghanistan-war-veteran-uniform.html (accessed 11 November 2010).

Daily Mail Reporter 2009. Ainsworth Plans Law to Let Soldiers Jump Waiting Lists, *Daily Mail* (online), 18 December. Available at: http://www.dailymail.co.uk/news/article-1236776/Defence-Secretary-Bob-Ainsworth-plans-law-let-soldiers-jump-waiting-lists.html (accessed 19 December 2009).

Daily Mail 2010. Tories to Tackle Mental Health 'Time Bomb' in Armed Forces, *Daily Mail* (online), 20 April. Available at: http://www.dailymail.co.uk/news/election/article-1268002/General-Election-2010-Tories-tackle-mental-health-time-bomb-armed-forces.html (accessed 14 May 2011).

Daily Telegraph 2010. Supreme Court Ruling: British Soldiers Abroad 'Not Protected By Human Rights Laws', *Telegraph* (online), 30 June. Available at: http://www.telegraph.co.uk/news/uknews/defence/7862765/Supreme-Court-ruling-British-soldiers-abroad-not-protected-by-human-rights-laws.html (accessed 1 July 2010).

Dandeker, Christopher 2001. On the Need to Be Different: Military Uniqueness and Civil–Military Relations in Modern Society, *RUSI Journal*, 126:3, pp. 4–9.

Dandeker, Christopher, Wessely, Simon, Iversen, Amy and Ross, John 2003. *Improving the Delivery of Cross Departmental Support and Services for Veterans*. Department of War Studies and the Institute of Psychiatry, King's College London.

Dannatt, General Sir Richard 2007. International Institute of Strategic Studies, London, 21 September. Available at: http://www.mod.uk/DefenceInternet/AboutDefence/People/Speeches/ChiefStaff/20070921 (accessed 14 May 2009).

—— 2010a. *Leading from the Front*. London: Bantam Press.

—— 2010b. Evidence to Chilcot Inquiry, 28 July.

—— 2012 (General Lord, KCB CBE MC). Interview with author, London.

Davies, Nick 2008. *Flat Earth News*. London: Chatto & Windus.

Davies of Stamford, Lord (Quentin) 2012. Interview with author, London.

Dawson, Matt 2012. Against the Big Society: A Durkheimian Socialist Critique, *Critical Social Policy*. Available at: http://csp.sagepub.com/content/early/2012/02/22/0261018312439364.abstract (accessed 15 September 2012).

De Watteville, Herman 1954. *The British Soldier: His Daily Life from Tudor to Modern Times*. London: Dent.

Defence Analytical Services Agency 2010. *Strength of the Trained UK Regular Forces By Service and Nationality*, Table 2.14. Available at: http://www.dasa.mod.uk/modintranet/UKDS/UKDS2010/c2/table214.php (accessed 21 November 2011).

Defence Select Committee 2000–01. *Strategic Defence Review: Evidence and Appendices* HC 29-II.

—— 2000–01. *Strategic Defence Review: Policy for People Report* 2000–01 HC 462.

—— 2000–01. *Strategic Defence Review: Report and Proceedings of the Committee* HC 29-I.
—— 2002–03. *A New Chapter to the Strategic Defence Review* HC 93-I.
—— 2003–04. *Lessons of Iraq* HC 57-I.
—— 2004–05. *Duty of Care* HC 63-1 and HC 63-II.
—— 2004–05. *Iraq: An Initial Assessment of Post-Conflict Operations* HC 65-I and HC 65-II.
—— 2005–06. *Annual Report and Accounts 2004–05* HC1293.
—— 2005–06. *Costs of Peace-Keeping in Iraq and Afghanistan: Spring Supplementary Estimate* HC 980.
—— 2005–06. *Educating Service Children* HC 1054.
—— 2005–06. *The UK Deployment to Afghanistan* HC 558.
—— 2005–06. *UK Operations in Iraq* HC 1241.
—— 2006–07. *Ministry of Defence Annual Report and Accounts 2005–2006* HC 57.
—— 2006–07. *The Army's Requirement for Armoured Vehicles: The FRES Programme* HC 159.
—— 2006–07. *UK Operations in Afghanistan* HC 408.
—— 2007–08. *Medical Care for the Armed Forces* HC 327.
—— 2007–08. *Recruiting and Retaining Armed Forces Personnel* HC 424.
—— 2008–09. *Helicopter Capability* HC 434.
—— 2010–12. *Operations in Afghanistan* HC 554.
Department of Defense 2002. *A New Social Compact: A Reciprocal Partnership Between the Department of Defense, Service Members and Families.* Available at: www.militaryhomefront.dod.mil/dav/lsn/LSN/BINARY_RESOURCE/BINARY_CONTENT/1665301 (accessed 6 June 2010).
Desch, Michael 2007. Bush and the Generals, *Foreign Affairs*, 83:3, pp. 97–108.
Development, Concepts and Doctrine and Centre 2010. *Global Strategic Trends – Out to 2040*, 4th edition.
Dixon, Norman 1976. *On the Psychology of Military Incompetence.* London: Pimlico.
Dobbin, Rev Dr Victor, CB MBE 2012. Discussion with author.
Donaldson, Rt Hon Jeffrey MP 2012. Interview with author, London.
Done, Francis CBE 2012. Interview with author, London.
Donnelly, Tom (Moderator) 2006, Round Table Discussion, Rummy and His Generals, *Armed Forces Journal.* Available at: http://www.armedforcesjournal.com/2006/06/1813589/ (accessed 19 December 2011).
Dorman, Andrew M. 2006. Reorganising the Infantry: Drivers of Change and What This Tells Us about the State of the Defence Debate Today, *British Journal of Politics and International Relations*, 8:4, November, pp. 489–502.
—— 2011. Making 2 + 2 = 5, The 2010 Strategic Defence and Security Review, *Defence and Security Analysis*, 27:1, pp. 77–87.
Downs, Anthony 1972. Up and Down with Ecology: the Cycle, *Public Interest*, 28, Summer, pp. 38–49.

Doyle, Jack 2011. The Men Who Disgraced the Army, *Daily Mail*, 8 September, p. 13.

Drayson, Paul (Lord) 2010. Note to Chilcot Inquiry, 15 December. Available at: http://www.iraqinquiry.org.uk/media/50260/LordDraysonofKensington-statement.pdf (accessed 13 November 2011).

Drewry Gavin 2005. Citizens as Customers: Charters and the Contractualization of Quality in Public Services, EGPA Conference, Bern. Available at: http://public-admin.co.uk/brochures/cutomers_charters_paper.pdf (accessed 2 July 2011).

Drury, Ian 2010. Ministers Were Warned FOUR YEARS Ago that Snatch Land Rovers Need Replacing, *Daily Mail* (online), 27 July. Available at: http://www.dailymail.co.uk/news/article-1298076/Ministers-warned-FOUR-YEARS-ago-Snatch-Land-Rovers-needed-replacing.html (accessed 27 July 2010).

Durkheim, Emile 1984. *The Division of Labour in Society* (introduced by Lewis Coser, trans. W.D. Halls). Basingstoke: Macmillan Press.

East Suffolk Gazette 1863–1866, National Army Museum, London SW3, 27826.

The Economist 2005. The Good Company, *The Economist* (online), 20 January. Available at: http://www.economist.com/node/3555212 (accessed 2 November 2011).

—— 2009. Overstretched, Overwhelmed and Over There, *The Economist* (online), 12 January. Available at: http://www.economist.com/node/13022029 (accessed 27 December 2010).

—— 2009. Losing their Way? *The Economist* (online), 29 January. Available at: http://www.economist.com/node/13022177 (accessed 27 December 2010).

Edmonds, Timothy and Forster, Anthony 2007. *Out of Step: The Case for Change in the British Armed Forces*. London: Demos.

Edmunds, Timothy 2012. British Civil–Military Relations and the Problem of Risk, *International Affairs*, 88:2, pp. 265–282.

Elazar, Daniel J. 1996. *Covenant and Commonwealth: From Christian Separation Through the Protestant Reformation, the Covenant Tradition in Politics*, Vol. 2. New York: Transaction Publishers.

Elliott, Mark 2010. Pretending You Can Make an Aspiration into Law, 12(7), *Parliamentary Brief*, March.

England Footballers Foundation. Available at: http://www.englandfootballersfoundation.com/index.html (accessed 15 June 2011).

Epili Uluilakeba v Secretary of State May 2011. Immigration and Asylum, First Tier Tribunal. (acquired from private sources).

Etherington, Mark CBE 2010. Evidence to Chilcot Inquiry, 9 July.

Evans, Michael 2000. Army Covenant Tells Soldiers to Expect Sacrifice, *The Times*, 3 April.

—— 2005. How British Army is Fast Becoming A Foreign Legion, *The Times*, 14 November, p. 7.

—— 2007. Forces' Housing Criticised, *The Times*, 4 January, p. 8.

—— 2007. MoD Rebuffs General's Claim that Troops Are Devalued and Angry, *The Times*, 19 November, p. 14.

—— 2009. Troops Get Longer Breaks to Save Marriages, *The Times*, 20 January, p. 22.
—— 2009. Britain Opens Door to Thousands More Gurkhas After Policy U-Turn, *The Times*, 29 January, p. 17.
Farrow, Malcolm 2011. Interview with author, London.
Feaver, Peter D. 1996. The Civil–Military Problematique: Huntington, Janowitz and the Question of Civilian Control, *Armed Forces and Society*, 23:2, Winter, pp. 149–178.
—— 2003a The Civil–Military Gap in Comparative Perspective, *Journal of Strategic Studies*, 26:2, pp. 1–5.
—— 2003b. *Armed Servants: Agency, Oversight and Civil–Military Relations.* Cambridge, MA: Harvard University Press.
Feaver, Peter D. and Kohn, Richard H. (eds) 2001. *Soldiers and Civilians: The Civil–Military Gap and American National Security*, Triangle Institute of Security Studies, Belfer Center for Science and International Affairs. Cambridge, MA: MIT Press.
Ferguson, Niall 2001. *Empire: How Britain Made the Modern World.* London: Allen Lane.
Fergusson, James 2009. *A Million Bullets: The Real Story of the British Army in Afghanistan.* London: Corgi.
—— 2011. Correspondence with author.
—— 2011. *Taliban: The True Story of the World's Most Feared Guerrilla Fighters.* London: Corgi.
Fielding, Helen 1996. *Bridget Jones's Diary.* London: Picador.
Finer, Samuel E. 2002 (Introduction, Jay Stanley). *The Man on Horseback: The Role of the Military in Politics.* New Jersey: Transaction Publishers (originally published Pall Mall, London, 1962).
Fisher, David 2011. *Morality and War: Can War Be Just in the Twenty-First Century?* Oxford: Oxford University Press.
Forster, Anthony 2005. *Armed Forces and Society in Europe.* Basingstoke: Palgrave Macmillan.
—— 2006. Breaking the Covenant: Governance of the British Army in the Twenty-First Century, *International Affairs*, 82:6, pp. 1043–1057.
—— 2011. Interview with author, Durham.
—— 2012a. The Military Covenant and British Civil–Military Relations, Letting the Genie out of the Bottle, *Armed Forces and Society*, 38:2, pp. 273–290.
—— 2012b. British Judicial Engagement and the Juridification of the Armed Forces, *International Affairs*, 88:2, pp. 283–300.
Fowler, Rebecca 1996. Soldiers Guilty of 'Horrific' Murder, *The Independent* (online), 29 March. Available at: http://www.independent.co.uk/news/soldiers-guilty-of-horrific-murder-1344601.html (accessed 12 April 2010).
Fox, Rt Hon Liam MP 2010. Letter to Prime Minister (Daily Telegraph, Defence Cuts: Liam Fox's Leaked Letter in Full, 28 September), p. 1.
—— 2010b. Speech, Conservative Party Conference, Birmingham.

Freedman, Lawrence 1999. *The Politics of British Defence 1979–1998.* Basingstoke: Palgrave Macmillan.

French, David 2002. Doctrine and Organisation in the British Army 1919–1932, *Historical Journal*, 44:2, pp. 417–515.

—— 2004. *Military Identities: The Regimental System, the British Army and the British People c 1870–2000.* Oxford: Oxford University Press.

French, David and Holden Reid, Brian (eds) 2002. *The British General Staff: Reform and Innovation c.1890–1939.* London: Frank Cass.

Fried, Charles 1981. *Contract as Promise: A Theory of Contractual Obligations.* Cambridge, MA and London, England: Harvard University Press.

Fry, Lt Gen Sir Robert KCB, CBE 2009. Evidence to Chilcot Inquiry, 16 December.

Fukuyama, Francis 1995. *Trust: The Social Virtues and the Creation of Prosperity.* London: Hamish Hamilton.

Fulton, Lt Gen Sir Robert, KBE 2010. Evidence to Chilcot Inquiry, 27 July.

Fussell, Paul 1975. *The Great War and Modern Memory.* Oxford: Oxford University Press (Anniversary Edition, 2000).

Gabriel, Richard A. and Metz, Karen S. 1992. *A History of Military Medicine, Vol II, From the Renaissance through Modern Times.* Westport and London: Greenwood Press.

Gage, Rt. Hon. Sir William 2011. *The Report of the Baha Mousa Inquiry*, HC 1452.

Gambetta, Diego (ed.) 1998. *Trust: Making and Breaking Cooperative Relations.* Oxford: Basil Blackwell.

Gaunt, Jon 2009. Gurkhas Not Shirkers, *The Sun*, 1 May, p. 37.

Gerber, David A. (ed.) 2000. *Disabled Veterans in History.* Ann Arbor: University of Michigan Press.

Gibbon, Edward, reprinted edition 1984. *Memoirs of My Life.* Harmondsworth: Penguin.

Giddens, Anthony 1971. *Capitalism and Modern Social Theory: an Analysis of the Writings of Marx, Durkheim and Max Weber.* Cambridge: Cambridge University Press.

—— 1991. *Modernity and Self-Identity: Self and Society in the Late Modern Age.* Cambridge: Polity Press.

Gillan, Audrey 2007. Battle of Housing for Injured Troops' Families, *The Guardian* (online), 21 July. Available at: http://www.guardian.co.uk/uk/2007/jul/31/military.immigrationpolicy (accessed 23 January 2011).

—— 2007. We've Been Neglected and Let Down Say Combat Troops, *The Guardian* (online), 15 August. Available at: http://www.guardian.co.uk/uk/2007/aug/15/military.iraq1 (accessed 23 January 2011).

Gioia, J.L. and Herman, R.E. 2001. Family-Centered Employment – A Survey of Current Practices, The Herman Group (Social Compact Supplementary Paper).

Gladwell, Malcolm 2000. *The Tipping Point: How Little Things Can Make a Big Difference.* London: Abacus.

Glasman, Maurice 2010. Society Not State: The Challenge of the Big Society, *Public Policy Research*, 17:2, pp. 59–64.

Glover, Julian, Norton-Taylor, Richard and Wintour, Patrick 2006. Iraq: Voters Want British Troops Home By the End of the Year, *The Guardian* (online), 24 October. Available at: http://www.guardian.co.uk/politics/2006/oct/24/uk.iraq (accessed 25 January 2011).

Goffman, Erving 1990. *The Representation of Self in Everyday Life*. Harmondsworth: Penguin.

Gouldner, Alvin W. 1960. The Norm of Reciprocity: A Preliminary Statement, *American Sociological Review*, 25:2, pp. 65–70.

Gower, Melanie 2011. *Naturalisation as a British Citizen*, House of Commons Standard Note SN/HA/3232.

Granovetter, Mark 1978. Threshold Models of Collective Behaviour, *American Journal of Sociology*, 83:6, May, pp. 1420–1443.

Gray, Bernard 2009. *Review of Acquisition for the Secretary of State for Defence*. Available at: http://webarchive.nationalarchives.gov.uk/20120913104443/http://www.mod.uk/NR/rdonlyres/78821960–14A0–429E-A90AFA2A8C292C84/0/ReviewAcquisitionGrayreport.pdf (accessed 15 December 2011).

Gray, James MP 2012. Interview with author, London.

Greenpeace, Greenwashing (ongoing). Available at: http://stopgreenwash.org/ (accessed 13 October 2011).

Gregory, Adrian 1994. *The Silence of Memory: Armistice Day 1919–1946*. Oxford: Berg.

Grey, Stephen 2009. *Retreat from Basra – Learning the Lessons*, StephenGrey.com, 20 September.

—— 2010. *Operation Snakebite: The Explosive True Story of an Afghan Desert Siege*. Harmondsworth: Penguin.

The Guardian Leader 2007. Military Covenant: Poppy Politics, *The Guardian* (online), 2 November. Available at: http://www.guardian.co.uk/commentisfree/2007/nov/03/politics.military (accessed 23 January 2011).

The Guardian, Guardian Data Blog (ongoing at time of writing). *British Dead and Wounded in Afghanistan, Month by Month*. Available at: http://www.guardian.co.uk/news/datablog/2009/sep/17/afghanistan-casualties-dead-wounded-british-data?commentpage=all#start-of-comments (ongoing at time of writing).

The Guardian, Media Guardian 2010. National Daily Newspaper Circulation *The Guardian* (online). Available at: http://media.guardian.co.uk/presspublishing/table/0,2191687,00.html (accessed 28 May 2010).

Guest, David 1998. Is the Psychological Contract Worth Taking Seriously? *Journal of Organisational Behaviour*, 19, pp. 649–664.

Guest, David and Conway, Neil 2002. Communicating the Psychological Contract: An Employer Perspective, *Human Resource Management Journal*, 12:2, pp. 22–38.

Guthrie, Charles (General Sir Charles) 2000. *Chief of the Defence Staff's Annual Lecture*, RUSI, London, December.

—— (Field Marshal Lord Guthrie of Craigiebank, GCB LVO OBE DL) 2011. Interview with author, London.

Guthrie, Charles and Quinlan, Michael 2007. *Just War: The Just War Tradition, Ethics in Modern Warfare.* London: Bloomsbury.

Hackett, General Sir John 1983. *The Profession of Arms.* London: Sidgwick and Jackson.

Haddon-Cave, Charles QC 2009. *The Nimrod Review: an independent review into the broader issues surrounding the loss of the RAF Nimrod MR2 aircraft XV230 in Afghanistan in 2006.* London: TSO HC1025.

Hahn, Scott W. 2009. *Kinship by Covenant, Covenant Research in Modern Biblical Scholarship.* New Haven: Yale University Press.

Hardin, Garrett 1968. The Tragedy of the Commons, *Science*, 162, pp. 1243–1248.

Hardin, Russell 2006. *Trust.* Cambridge: Polity Press.

Harding, Thomas 2006. Muslim Accosts Injured Para in Hospital, *Daily Telegraph*, 2 October, p. 1.

—— 2007. Wounded Soldiers 'Suffering MoD Secrecy', *Daily Telegraph* (online), 17 July. Available at: http://www.telegraph.co.uk/news/uknews/1524117/Wounded-soldiers-suffering-MoD-secrecy.html (accessed 14 February 2011).

—— 2009. SAS Chief Quits Over Negligence that Killed His Troops, *Daily Telegraph* (online), 7 March. Available at: http://www.telegraph.co.uk/news/newstopics/onthefrontline/3332417/Exclusive-SAS-chief-quits-over-negligence-that-killed-his-troops.html (accessed 8 August 2010).

—— 2009. MoD Penny-Pinching Cost 14 Lives, *Daily Telegraph* (online), 29 October. Available at: http://www.telegraph.co.uk/news/uknews/3343653/Muslim-accosts-injured-Para-in-hospital.html (accessed 29 October 2009).

Hardy, E.J. 1900. *Mr Thomas Atkins.* London: T Fisher Unwin.

Harmen Akenson, Donald 1992. *God's Peoples: Covenant and Land in South Africa, Israel and Ulster.* Ithaca and London: Cornell University Press.

Harnden, Toby 2011. *Dead Men Risen.* London: Quercus.

Harpin, Lee 2011. 10,000 Ex-Soldiers Will Be Homeless this Christmas, *The People* (online), 18 December. Available at: http://www.people.co.uk/news/uk-world-news/2011/12/18/10–000-ex-soldiers-will-be-homeless-this-christmas-102039–23643084/ (accessed 18 December 2011).

Hastings, Max 2006. Our Armed Forces Have Allowed Themselves To be Cowered into Silence, *The Guardian* (online), 9 October. Available at: http://www.guardian.co.uk/commentisfree/2006/oct/09/comment.afghanistan (accessed 8 August 2010).

—— 2007. Despite Public Anger, the Army Still See Afghanistan as a Cause Worth Dying For, *The Guardian* (online), 28 August. Available at: http://www.guardian.co.uk/commentisfree/2007/aug/28/comment.afghanistan (accessed 15 January 2011).

—— 2008. A Land Unfit For Heroes, *Daily Mail* (online), 8 March. Available at: http://www.dailymail.co.uk/debate/columnists/article-528633/A-land-unfit-heroes.html (accessed 10 July 2010).

Haynes, Deborah, Loyd, Antony, Kiley, Sam and Coghlan, Tom 2010. Officers' Mess: They Went into Helmand with their Eyes Shut and their Fingers Crossed, *The Times*, 9 June, p. 1.

Hennessy, Patrick, Rayment, Sean and Amoore, Miles 2009. Parents of Soldiers Killed in Afghanistan Lash Out at Government, *Daily Telegraph* (online), 11 July. Available at: http://www.telegraph.co.uk/news/newstopics/onthefrontline/5806043/Parents-of-soldiers-killed-in-Afghanistan-lash-out-at-Government.html (accessed 4 July 2010).

Herman, Edward S. and Chomsky, Noam 1994. *Manufacturing Consent: The Political Economy of the Mass Media.* London: Vintage.

Hickley, Matthew 2006. How Troops Wounded in Action Don't Count with the MoD, *Daily Mail*, 17 January, p. 17.

Hickman, Tom 2004. *The Call Up: A History of National Service.* London: Headline.

Hobsbawm, Eric and Ranger, Terence (eds) 1989. *The Invention of Tradition.* London: Canto.

Holden Reid, Brian (ed.) 1993. *The Science of War: Back to First Principles.* Abingdon: Routledge.

—— 1995. *British Style of Command.* Annex A, Army Doctrine Publication, 1995, Volume 2, *Command*, AC 71564.

—— 1998. *Studies in British Military Thought: Debates with Fuller and Liddell Hart.* Lincoln, NE: University of Nebraska Press.

Holloway, Adam MP 2009. *The Failure of British Political and Military Leadership in Iraq.* London: First Defence. Available at: http://www.defenceviewpoints.co.uk/articles-and-analysis/the-failure-of-british-political-and-military-leadership-in-iraq (accessed 23 June 2010).

—— 2012. Interview with author, London.

Holmes, Richard 2001. *Redcoat: The British Soldier in the Age of Horse and Musket*. London: HarperCollins.

—— 2006. *Soldiers and Society.* The Annual Liddell Hart Lecture, King's College London.

—— 2011. *Soldiers: Army Lives and Loyalties From Redcoats to Dusty Warriors.* London: Harper Press.

Honeyman, Victoria 2009. Gordon Brown and International Policy, *Policy Studies*, 30:1, pp. 85–100.

Hoon, Rt Hon Geoffrey 2010. Evidence to Chilcot Inquiry, 19 January.

Houghton, Lt-Gen Sir Nicholas, GCB CBE 2010. Evidence to Chilcot Inquiry, 5 January.

Howard, Martin 2002. *Wellington's Doctors: British Army Medical Services in the Napoleonic Wars.* Staplehurst: Spellmount.

Howard, Michael (ed.) 1957. *Soldiers and Governments: Nine Studies in Civil–Military Relations.* London: Eyre and Spottiswoode.

—— 1992. *The Lessons of History*. New Haven: Yale University Press.

Huntington, Samuel 1957. *The Soldier and the State*. Harvard: Harvard University Press.

ICM/Sunday Telegraph 2008. Poll, 12/13 March. Available at: http://www.icmresearch.com/pdfs/2008_mar_sun_tele_iraq_poll.pdf (accessed 15 January 2011).

The Independent Editorial 2009. In Breach of the Military Covenant, *The Independent* (online), 28 February, p. 44.

The Independent on Sunday 2007. Britain Must Honour its Commitment to Our Troops, *The Independent on Sunday* (online), 2 September. Available at: http://www.independent.co.uk/opinion/leading-articles/leading-article-britain-must-honour-its-commitment-to-our-troops-401142.html (accessed 3 July 2010).

Ingham, Sarah and Dandeker, Christopher 2010. The Covenant We Must Protect From the Lawyers, *Parliamentary Brief*, June.

Ingram, Rt. Hon. Adam Ingram 2011. Interview with author, London.

Ipsos MORI 2008. War with Iraq: Trends 2002–2007. Available at: http://www.internationalresearch.co.uk/researchpublications/researcharchive/poll.aspx (accessed 15 January 2011).

Ipsos MORI 2009. Issues Index, April. Available at: http://www.ipsos-mori.com/Assets/Docs/Polls/issues-index-april-2009-charts.pdf (accessed 15 January 2011).

Ipsos MORI 2009. Two-Minute Silence, 9 November. Available at: http://www.internationalresearch.co.uk/researchpublications/researcharchive/2510/Two-Minute-Silence-as-important-as-ever-say-Britons.aspx (accessed 15 January 2011).

Jackson, General Sir Mike GCB CBE DSO DL 2005. *Declassified Extracts From a Report of a Visit to Iraq.* Available at: http://www.iraqinquiry.org.uk/media/48836/smj-report-oct-05.pdf (accessed 4 April 2012).

—— 2006. BBC: The Dimbleby Lecture, London. Available at: http://www.bbc.co.uk/pressoffice/pressreleases/stories/2006/12_december/07/dimbleby.shtml (accessed 14 May 2009).

—— 2008. *Soldier: The Autobiography.* London: Corgi.

—— 2010. Evidence to Chilcot Inquiry, 28 July.

Janowitz, Morris 1960. *The Professional Soldier: A Social and Political Portrait.* New York: The Free Press (Macmillan).

Jenkins, Simon 2006. A Simple Minded General Stirs Mutiny in the Ranks, *Sunday Times*, 15 October, p. 16.

Jenks, Timothy T. 2006. *Naval Engagements: Patriotism, Cultural Politics and the Royal Navy, 1793–1815.* Oxford: Oxford University Press.

Johnson, Boris 2007. How, as Mayor, I would help our brave Troops, *The Spectator*, 13 December 2007.

Johnstone-Burt, Vice Admiral Tony, MBE 2012. Discussion and correspondence with author.

Joint Doctrine Publication 1997. *British Defence Doctrine*, JWP0–01.

—— 2008. *British Defence Doctrine*, JWP 0–01, 3rd edition.

Jolly, Schona 2010. Human Rights Still Matter on the Battlefield, *The Guardian*, 30 June. Available at: http://www.guardian.co.uk/commentisfree/libertycentral/2010/jun/30/human-rights-act-soldiers-battlefield (accessed 30 June 2010).

Jones, Edgar and Wessely, Simon 2005. *Shellshock to PTSD: Military Psychiatry from 1900 to the Gulf War.* Maudsley Monographs, Psychology Press, Hove: Taylor and Francis Group.

Jones, Harry W. 1975. The Jurisprudence of Contracts, *The Cincinnati Law Review*, 54, pp. 43–57.

Jones, Kevan MP 2012. Interview with author, London.

Journal of the Household Brigade, National Army Museum Collection, London SW3, 15462.

Joyce, Eric 1997. *Armed and the Man: Renewing the Armed Services.* London: Fabian Society.

Judd, Terri, Goodchild, Sophie, Johnson, Andrew, Veevers Lauren and Sengupta, Kim 2007. The Betrayal of Britain's Fighting Men and Women, *The Independent on Sunday*, 11 March, p. 14.

Kaldor, Mary 2007. *New and Old Wars: Organised Violence in a Global Era.* Stanford: Stanford University Press.

Kampfner, John 2004. *Blair's Wars.* London: Free Press.

Kane, Patricia 2005. Facing Death in Iraq for £3.60 an hour, *Mail on Sunday*, 27 February, p. 15.

Kay, John 1993. *Foundations of Corporate Success.* Oxford: Oxford University Press.

—— 2010. Think Before You Tear Up an Unwritten Contract, *Financial Times*, 16 March. Available at: http://www.ft.com/cms/s/0/cc712f1a-3134–11df-8e6f-00144feabdc0.html#axzz218tkXWZ (accessed 16 March 2010).

Kay, John 2007. Do Your Duty By Our Brave Boys, *The Sun*, 17 September. Available at: http://www.thesun.co.uk/sol/homepage/news/294003/Do-your-duty-by-our-brave-boys.html (accessed 28 November 2011).

Keegan, John 2004. *The Iraq War: From Victory in 21 Days to the Insurgent Aftermath.* New York: Alfred A Knopf.

Kemp, Ross 2008. *Ross Kemp in Afghanistan.* DVD, Tiger Aspect/Mongoose.

—— 2009. *Return to Afghanistan.* DVD, Tiger Aspect/Mongoose.

Kerr, Rachel 2008. A Force for Good? War, Crime and Legitimacy: The British Army in Iraq, *Defense and Security Analysis*, 24:4, pp. 401–419.

King, Alex 1998. *Memorials of the Great War in Britain: The Symbolism and Politics of Remembrance.* Oxford: Berg.

King, Anthony 2009. Colonel Iron and The Charge of the Knights, *Prospect*, March.

—— 2010a. The Afghan War and Postmodern Memory: Commemoration and the Dead of Helmand, *British Journal of Sociology*, 1, pp. 1–25.

—— 2010b. Understanding the Helmand Campaign: British Military Operations In Afghanistan, *International Affairs*, 86:2, March, pp. 311–332.

King's Forum 1998. *The SDR: How Strategic? How Much of a Review?* London Defence Studies, London: Brassey's.

Kohn, Richard H. 2009. Always Salute, Never Resign, *Foreign Affairs*, November.

Krebs, Ronald R. 2009. The Citizen-Soldier Tradition in the United States: Has Its Demise Been Greatly Exaggerated? *Armed Forces and Society*, 36:1, October, pp. 153–174.

Labour Party 2010. *General Election Manifesto*. Available at: http://www.labour.org.uk/uploads/TheLabourPartyManifesto-2010.pdf (accessed 6 June 2011).

Lea, Michael 2008. Hurt Heroes 'Made to Grovel', *The Sun* (online), 5 March. Available at: http://www.thesun.co.uk/sol/homepage/news/876107/Falklands-hero-Simon-Weston-attack-on-ministers-Armed-Forces-Wounded-heroes-made-to-grovel.html (accessed 21 November 2011).

Leake, Christopher 2008. Defence Chiefs Rocked by a Crisis in Morale After MoD Surveys 10,000 Troops, *Daily Mail*, 9 August. Available at: http://www.dailymail.co.uk/news/article-1205286/Defence-chiefs-rocked-crisis-morale-MoD-surveys-10-000-troops.html (accessed 5 January 2011).

Ledwidge, Frank 2011. *Losing Small Wars: British Military Failure in Iraq and Afghanistan.* New Haven and London: Yale University Press.

Lee, Robert 2011. Interview with author, London.

Lewis Hudson, Geoffrey 1988. *The English Privy Council and the Relief of Disabled Soldiers, c. 1558–1625.* Unpublished thesis. Available at: http://digitalcommons.mcmaster.ca/opendissertations/4108/ (accessed 1 September 2011).

Lewis, J. David and Weighart, Andrew 1985. Trust as a Social Reality, *Social Forces*, 63:4, pp. 967–986.

Liberal Democrat Party 2010. *General Election Manifesto.* Available at: http://www.libdems.org.uk/our_manifesto.aspx (accessed 6 June 2011).

Liddell Hart, Basil 1932. *The British Way in Warfare.* London: Faber.

Liebenstein, H. 1950. Bandwagon, Snob and Veblen Effects in the Theory of Consumers' Demand, *Quarterly Journal of Economics*, 64:2, May, pp. 183–207.

Limbu & Ors, R (on the application of) v Secretary of State for the Home Department & Ors 2008. EWHC 2261 (Admin) (30 September 2008). Available at: http://www.bailii.org/ew/cases/EWHC/Admin/2008/2261.html (accessed 3 November 2011).

Lloyd, Rt Hon Lord of Berwick 2004. *The Lloyd Inquiry: Independent Public Inquiry into Gulf War Illnesses.* East Grinstead: Alpha Office. Available at: http://www.lloyd-gwii.com/admin/ManagedFiles/4/LloydReport.pdf (accessed 15 May 2012).

Lord Ashcroft Polls 2012. *Armed Forces and Society.* Available at: http://lordashcroftpolls.com/2012/05/the-armed-forces-society/ (accessed 12 May 2012).

Macaulay, Stewart 1963. Non-Contractual Relations in Business: A Preliminary Study, *American Sociological Review*, 28:1, pp. 55–67.

MacDermid, S.M., Weiss, H.M., Beal, D.J. and Strauss, R. 2001. Social and Psychological Contracts: Relations Between Workers and Employers in the Twentieth Century, *Social Compact Supplementary Paper B*, Military Families Research Institute, Purdue.

MacInnes, Colin 2007. The British Army's New Way in Warfare: A Doctrinal Misstep? *Defense & Security Analysis*, 23:2, June, pp. 127–141.

Macintyre, Alasdair 1981. *After Virtue*. London: Bristol Classic Press (3rd edition).

MacKenzie, John M. (ed.) 1992. *Popular Imperialism and the Military*. Manchester: Manchester University Press.

Mackinlay, W.G.L. 2007. Perceptions and Misconceptions: How are International and UK Law Perceived to Affect Military Commanders and their Subordinates on Operations? *Defence Studies*, 7:1, pp. 111–160.

Macpherson, Sir Nicholas, KCB 2010. Evidence to Chilcot Inquiry, 22 January.

Maher, Shiraz 2011. *Ties that Bind: How the story of Britain's Muslim Soldiers Can Forge a National Identity*. London: Policy Exchange.

Mangan J.A. and Walvin, James (eds) 1987. *Manliness and Morality: Middle-Class Masculinity in Britain and America*. Manchester: Manchester University Press.

McCarthy, Julie 2012. Interview with author, London.

McCartney, Helen 2010. The Military Covenant and the Civil–Military Contract in Britain, *International Affairs*, 86:2, pp. 411–428.

—— 2011. Hero, Victim or Villain? The Public Image of the British Soldier and its Implications for Defence Policy, *Defense and Security Analysis*, 27:1, pp. 43–54.

McCombs, Maxwell 2004. *Setting the Agenda: The Mass Media and Public Opinion*. Cambridge: Polity Press.

McCombs, Maxwell and Shaw, Donald L. 1972. The Agenda Setting Function of the Mass Media, *Public Opinion Quarterly*, 36:2, pp. 176–187.

McNab, Andy 2008. Andy McNab on the Battle that Never Ends, *Daily Telegraph*, 22 November. Available at: http://www.telegraph.co.uk/news/telegraphchristmasappeal/3503105/The-battle-that-never-ends.html (accessed 27 November 2011).

Mercer, Patrick OBE MP 2011. Interview with author, London.

Metz, Steven 1994. A Wake for Clausewitz, *Parameters*, Winter 1994–95, pp. 126–132.

Mileham, Patrick 1998. Military Virtues 1, The Right to be Different? *Defence & Security Analysis*, 14:2, pp. 169–189.

—— 2010. Unlimited Liability and the Military Covenant, *Journal of Military Ethics*, 9:1, pp. 23–40.

Milroy, Dr Hugh, OBE 2011. Interview with author, London.

Ministry of Defence (MoD) 1998. *The Strategic Defence Review* CM 3999.

—— 2000. *Armed Forces Overarching Personnel Strategy*. Available at: http://www.mod.uk/issues/afops/index.htm (accessed 3 October 2009).

—— 2002. *The Strategic Defence Review: A New Chapter* CM5566.

—— 2003a *Delivering Security in a Changing World: Future Capabilities* CM 6041-I.

—— 2003b. *Delivering Security in a Changing World: Supporting Essays* CM6041-II.

—— 2007 (ongoing). *Armed Forces Continuous Attitude Survey.* Available at:http://www.dasa.mod.uk/index.php/publications/other/surveys/armed-forces-continuous-attitude-survey (accessed 15 March 2010).
—— 2007. Government Response to the RBL's *Honour the Covenant* Campaign October.
—— 2008a. *Report of Inquiry into National Recognition of Our Armed Forces* (The Davies Report). Available at: http://www.ppu.org.uk/militarism/recognition_of_our_armed_forces.pdf (accessed 4 June 2010)/
—— 2008b. Service Personnel Command Paper *The Nation's Commitment: Cross-Government Support to our Armed Forces, their Families and Veterans* CM7424 TSO. Available at: http://www.officialdocuments.gov.uk/document/cm74/7424/7424.pdf (accessed 4 June 2010).
—— 2008c. *Military Covenant Factsheet.*
—— 2010a. *Adaptability and Partnership: Issues for the Strategic Defence Review* CM 7794.
—— 2010b. *The Report of the Military Covenant Task Force.* Available at: https://www.gov.uk/government/uploads/system/uploads/attachment_data/file/27393/militarycovenanttaskforcerpt.pdf (accessed 30 September 2010).
—— 2010c. *The Government's Response to the Report of the Task Force on the Military Covenant.* Available at: https://www.gov.uk/government/uploads/system/uploads/attachment_data/file/27394/Armed_Forces_Covenant_Taskforce_Government_response.pdf (accessed 11 December 2010).
—— 2010d. *Defence Spending: Key Facts about Defence.*
—— 2011a. *The Armed Forces Covenant.* Available at: https://www.gov.uk/government/uploads/system/uploads/attachment_data/file/49469/the_armed_forces_covenant.pdf (accessed 30 May 2011).
—— 2011b. *The Armed Forces Covenant: Today and Tomorrow.* Available at: https://www.gov.uk/government/uploads/system/uploads/attachment_data/file/49470/the_armed_forces_covenant_today_and_tomorrow.pdf (accessed 30 May 2011).
—— 2011c. *Chief of the General Staff responds to the Baha Mousa Inquiry Report*, Defence Policy and Business News, 8 September. Available at: https://www.gov.uk/government/news/chief-of-the-general-staff-responds-to-baha-mousa-inquiry-report (accessed 9 September 2011).
—— 2011d. *Future Reserves 2020: Independent Commission to Review the United Kingdom's Reserve Forces* TSO. Available at: https://www.gov.uk/government/uploads/system/uploads/attachment_data/file/28394/ (accessed 30 July 2011).
Morris, Jonathan S. and Forgette, Richard 2007. News Grazers, TV News, Political Knowledge and Engagement, *Harvard International Journal of Press/Politics*, 12, Winter, pp. 91–107.
Morris, Nigel and Sengupta, Kim 2009. Voters Turn Against War in Afghanistan, *The Independent*, 29 July. Available at: http://www.independent.co.uk/news/uk/politics/voters-turn-against-war-in-afghanistan-1763227.html (accessed 6 June 2010).

Mostyn, Brigadier (Retired) Philip 2012. Telephone interview with author, London.

Muir, Rick and Wetherell, Margaret 2010, April. *Identity, Politics and Public Policy*, London: Institute of Public Policy Research. Available at: http://www.open.ac.uk/socialsciences/identities/files/identity_politics_and_public_policy.pdf (accessed 4 April 2014).

Murrison, Dr Andrew, MP 2010. Interview with author, London.

—— 2011. *Tommy This an' Tommy That: The Military Covenant*. London: Biteback.

Myers, Richard B. and Kohn, Richard H. 2007. Salute and Disobey, *Foreign Affairs*, September–October.

Nagl, John 2005. *Learning to Eat Soup With a Knife: Counterinsurgency Lessons From Malaya and Vietnam*. Chicago and London: University of Chicago Press.

—— 2011. How the US Learned Counterinsurgency, RUSI Analysis, 2 September. Available at: http://www.rusi.org/analysis/commentary/ref%3AC4E60F10A9B878/#.UoZWuKVPqkQ (accessed 8 September 2011).

National Army Museum/ICM 2009. *Conflicts of Interest* survey. Available at: http://www.nam.ac.uk/press/national-army-museum-survey-reveals-uks-attitudes-towards-british-army-recent-conflicts (accessed 15 January 2011).

National Association of Probation Officers 2008. *Ex Armed Service Personnel in the Criminal Justice System*.

National Audit Office 2007–08. *Ministry of Defence, Chinook Mk 3 Helicopters* HC 512.

—— 2009–10. *Ministry of Defence Major Projects Report* HC85-II.

Neuberg, Victor E. 1989. *Gone for a Soldier: A History of Life in the British Ranks from 1642*. London: Cassell.

Newton Dunn, Tom 2007. From Helmand to Hellhole, *The Sun* (online), 26 October. Available at: http://www.thesun.co.uk/sol/homepage/news/387976/Army-unit-Return-to-England-Barracks-worse-than-jail.html (accessed 11 July 2010).

—— 2009. British Troops are Best Says US Leader David Patraeus, *The Sun* (online), 21 August. Available at: http://www.thesun.co.uk/sol/homepage/news/campaigns/our_boys/2599446/British-troops-are-best-says-US-leader-David-Petraeus.html (accessed 27 November 2011).

Nicol, Mark 2007. Muslim Women Abuse Soldiers at Troops' Hospital, *Mail on Sunday* (online), 10 June. Available at: http://www.dailymail.co.uk/news/article-461031/Muslim-women-abuse-soldier-troops-hospital.html (accessed 11 November 2010).

Nolan Committee 1994. *Summary of the Nolan Committee's First Report on Standards in Public Life*.

Norton-Taylor, Richard 2005. Survey Finds Most Soldiers Unhappy with Service Life, *The Guardian*, 5 November. Available at: http://www.guardian.co.uk/uk/2005/nov/05/military.richardnortontaylor (accessed 27 November 2011).

—— 2006. Britain's New Top Soldier: 'Can the Army Cope? I Say, Just', *The Guardian*, August. Available at: http://www.guardian.co.uk/uk/2006/sep/04/military.afghanistan (accessed 15 September 2009).

—— 2007. Soldiers Need More Support, Think Tank Warns Ministers, *The Guardian*, 5 November. Available at: http://www.guardian.co.uk/uk/2007/nov/05/thinktanks.military (accessed 13 November 2010).

—— 2007. Wounded Troops Deserve Better Treatment Says Legion, *The Guardian*, 13 September. Available at: http://www.guardian.co.uk/uk/2007/sep/13/military.richardnortontaylor (accessed 15 September 2009).

—— 2008. Defence Chief Welcomes New 'Duty of Care' Deal For Forces, *The Guardian*, 18 July. Available at: http://www.guardian.co.uk/politics/2008/jul/18/defence.military (accessed 16 May 2010).

—— 2010. Supreme Court Considers UK Soldiers' Right to Sue Over Military Missions, *The Guardian*, 15 March. Available at: http://www.guardian.co.uk/uk/2010/mar/15/soldiers-human-rights-supreme-court (accessed 15 March 2010).

—— 2011. Armed Forces Bill Dropped as Ministers 'Duck Covenant Pledge', *The Guardian*, 5 May. Available at: http://www.guardian.co.uk/uk/2011/may/05/military-covenant-bill-dropped-ministers-pledge (accessed 5 May 2011).

Norton-Taylor, Richard, Glover, Julian and Watt, Nicholas 2009. Public Support for War in Afghanistan is Firm, Despite Deaths, *The Guardian*, 13 July. Available at: http://www.guardian.co.uk/uk/2009/jul/13/afghanistan-war-poll-public-support (accessed 13 November 2010).

O'Donnell, Sir Gus, KCB 2011. Evidence to Chilcot Inquiry, 28 January.

O'Neill, Onora 2002. *A Question of Trust.* Cambridge: Cambridge University Press.

Olsen, Mancur 1971. *The Logic of Collective Action: Public Goods and the Theory of Groups.* Cambridge, MA: Harvard University Press (2nd edition).

Opinion Research Business/BBC Newsnight 2007. Poll of December 2007, cited by BBC, 14 December. Basra Residents Blame UK Troops. Available at: http://news.bbc.co.uk/1/hi/world/middle_east/7144437.stm (accessed 15 January 2011).

Osiel, Mark J. 2002. *Obeying Orders: Atrocity, Discipline and the Law of War.* New Brunswick and London: Transaction Publishers.

Oskamp, Stuart and Schultz, P. Wesley 2005. *Attitudes and Opinions.* Mahwah, NJ: Erlbaum (3rd edition).

Oxley, Geoffrey W. 1974. *Poor Relief in England and Wales 1601–1834.* Newton Abbott: David and Charles.

Page, Lewis 2007. *Lions, Donkeys and Dinosaurs.* London: Arrow Books.

Palazzo, Albert 2008. *From Moltke to Bin Laden: The Relevance of Doctrine in the Contemporary Military Environment*, Land Warfare Studies Centre, Canberra.

Parry, Bryn, OBE 2012. Interview with author, Tidworth.

Parsons, Tony 2009. Why We Must Honour Harry Patch and End this Tragic Waste of Human Life in Afghanistan, *Daily Mirror*, 1 August. Available at: http://www.mirror.co.uk/news/uk-news/why-we-must-honour-harry-patch-and-end-409926 (accessed 23 June 2011).

Plato 1987. *The Republic* (trans. Desmond Lee). Harmondsworth: Penguin.

Porter, Bernard 2006. *The Absent-Minded Imperialists: Empire, Society and Culture in Britain.* Oxford: Oxford University Press.

Porter, Patrick 2010. Last Charge of the Knights: Iraq, Afghanistan and the Special Relationship, *International Affairs*, 86:2, pp. 355–375.

Powell, Jonathan 2011. *The New Machiavelli: How to Wield Power in the Modern World.* London: Vintage Books.

Public Administration Select Committee 2010–11. *Who Does UK National Strategy?* HC 435.

Rawstorne, Tom 2007. How Labour is Failing to Honour its Promise to the Troops who Risk their Lives for the Nation, *Daily Mail* (online), 19 March. Available at:http://www.dailymail.co.uk/news/article-443138/How-Labour-failing-honour-promise-troops.html (accessed 15 April 2010).

Rayment, Sean 2007. 'Our Forces Can't Carry on Like This' says General Sir Richard Dannatt, *Daily Telegraph*, 18 November, p. 1.

Reid, Rt Hon Lord (John) of Cardowen 2010. Evidence to Chilcot Inquiry, 3 February.

Richards, David, Blunkett, David and Mathers, Helen 2008. Old and New Labour Narratives of Whitehall: Radicals, Reactionaries and Defenders of the Whitehall Model, *Political Quarterly*, 79:4, pp. 488–498.

Ricks, Thomas E. 2006. Army's Iraq Work Assailed by Briton, *Washington Post*, 11 January. Available at: http://www.washingtonpost.com/wpdyn/content/article/2006/01/10/AR2006011001456.html (accessed 1 December 2011).

—— 2007. *Making the Corps.* New York: Scribner.

Riddell, Mary 2009. Had We Listened to Harry Patch, We would Still Not Be in Helmand, *Daily Telegraph*, 27 July. Available at: http://www.telegraph.co.uk/comment/columnists/maryriddell/5919886/Had-we-listened-to-Harry-Patch-we-would-not-still-be-in-Helmand.html (accessed 23 June 2011).

Rifle Brigade Chronicle, National Army Museum Collection, London SW3, L/3045/63.

Ritchie, Major General (retired) Andrew, CBE 2011. Interview with author, London.

Robathan, Rt. Hon. Andrew, MP 2011. Interview with author, London.

Roberts, Major-General (retired) Sir Sebastian, KCVO MBE 2010. Interview with author, London.

Robinson, Paul, de Lee, Nigel and Carrick, Don 2006. *Ethics Education in the Military.* Aldershot: Ashgate.

Robinson, Sandra L. 1996. Trust and Breach of the Psychological Contract, *Administrative Science Quarterly*, 41, pp. 574–599.

Rodger, N.A.M. 1986. *The Wooden World: An Anatomy of the Georgian Navy.* London: Collins.

Rogers, Everett M. 2003. *Diffusion of Innovations.* New York: Free Press (5th edition).

Rousseau, D.M. 1989. Psychological and Implied Contracts in Organizations, *Employee Responsibilities and Rights Journal*, 2:2, pp. 121–39.

Rousseau, Jean Jacques 1968 (trans. and ed. Maurice Cranston). *The Social Contract.* Harmondsworth: Penguin Classics.

Royal British Legion 2007. *Honour the Covenant* Campaign literature (supplied by Frances Done).

—— 2010. *It's Time to do Your Bit.* Available at: http://eactivist.com/eacampaign/clientcampaign.do?ea.client.id=145&ea.campaign.id=4388 (accessed 1 May 2011).

Royal Engineers Officers' Widows Society. Available at http://www.army.mod.uk/royalengineers/association/638.aspx (accessed 23 June 2011).

Royal United Services Institute 2010. *Whither Welfare: Structuring Welfare in the Military Community.* Occasional Paper.

Royle, Trevor 1986. *The Best Years of Their Lives.* London: Michael Joseph.

Rubin, G.R. 2002. United Kingdom Military Law: Autonomy, Civilianization, Juridification, *Modern Law Review*, 65:1, pp. 36–57.

Ruger, Jan 2004. Nation, Empire and Navy: Identity Politics in the United Kingdom 1887–1914, *Past and Present*, 185, Nov, pp. 159–188.

Sacks, Jonathan 2002. *The Dignity of Difference.* London: Continuum.

Said, Edward 1994. *Culture and Imperialism.* London: Vintage.

Sands, Sarah 2006. A Very Honest General, *Daily Mail*, 13 October, p. 12.

Schelling, Thomas C. 2006. *Micromotives and Macrobehaviour.* New York and London: W.W. Norton (2nd edition).

Secretary to the Commissioners, Royal Hospital Chelsea 1872. *Papers Illustrative of the Origin and Early History of the Royal Hospital at Chelsea.* HMSO, London.

Seldon, Anthony 2009. *Trust: How We Lost it and How to Get it Back.* London: Biteback Publishing.

Seldon, Anthony and Lodge, Guy 2010. *Brown at Ten.* London: Biteback Publishing.

Shaw, Major-General Jonathan 2010. Evidence to Chilcot Inquiry, 11 January.

Shaw, Martin 1999. *Post Military Society: Militarism, Demilitarization and War at the End of the Twentieth Century.* Cambridge: Polity Press.

Sheffield, Gary 2005. *Doctrine and Command in the British Army: A Historical Overview.* Army Doctrine Publication, *Land Operations*, AC 71819.

Shipman, Tim 2006. We Must Quit Iraq Says New Head of the Army, *Daily Mail*, 13 October, p. 1.

—— 2010. SAS defied MoD to rescue two of its men held hostage in Iraq as top commanders 'prepared to quit' over ban on mission, *Daily Mail*, 6 May. Available at: http://www.dailymail.co.uk/news/article-1273507/SAS-defied-MoD-rescue-men-held-hostage-Iraq.html#ixzz1tNHlwnEK (accessed 28 April 12).

Shirreff, Lt Gen Sir Richard, KCB, CBE 2010. Evidence to Chilcot Inquiry, 11 January.

Simpkins, Chris 2011. Interview with author, London.

Skocpol, Theda 1992. *Protecting Mothers and Soldiers: The Political Origins of Social Policy in the United States.* Cambridge: Harvard University Press.

Smith, Anthony D. 1991. *National Identity.* Nevada: Nevada University Press.

Smith, Melvyn Charles 2008. *Awarded for Valour: A History of the Victoria Cross and the Evolution of British Heroism.* Basingstoke: Palgrave Macmillan.

Smith, Michael 2001. One in Three Troops in Slum Barracks, *Daily Telegraph*, 26 February. Available at: http://www.telegraph.co.uk/news/uknews/1324207/One-in-three-troops-lives-in-slum-barracks.html(accessed 13 December 2010).

—— 2005. Iraq Battle Stress Worse than WWII, *Sunday Times*, 6 November, p. 1.

Smith, Michael and Sparrow, Andrew 2004. Britain Spends Too Little on its Forces Says Poll, *Daily Telegraph* (online), 23rd August. Available at: http://www.telegraph.co.uk/news/uknews/1469981/Britain-spends-too-little-on-its-forces-says-poll.html (accessed 15 January 2011).

Smith, Rupert 2005. *The Utility of Force: The Art of War in the Modern World.* Harmondsworth: Penguin.

Snider, Don M. and Watkins, Gayle M.; Matthews, Lloyd J. (ed.), 2001. *The Future of the Army Profession.* NJ: McGraw-Hill Higher Education.

Snyder, Colonel Lewis R. 2009. *The Generals' Revolt and Civil–Military Relations*, US Army War College Paper. Available at: http://www.dtic.mil/cgi-bin/GetTRDoc?AD=ADA508321 (accessed 17 October 2011).

The Soldier's Catechism 1644. University of Toronto archive. Available at: http://archive.org/details/soldierscatechis00cromuoft (accessed 28 April 2012).

Spencer, Sir Peter, KCB 2010. Evidence to Chilcot Inquiry, 26 July.

Spiers, E.M. 1980. *The Army and Society 1815–1914.* London: Longman.

Stone, Lawrence (ed.) 1994. *An Imperial State at War: Britain from 1689–1815.* London: Routledge.

Strachan, Hew 1984. *Wellington's Legacy: The Reform of the British Army 1830–1854.* Manchester: Manchester University Press.

—— 1997. *The Politics of the British Army.* Oxford: Oxford University Press.

—— 2000 (ed.). *The British Army, Manpower and Society into the Twenty First Century.* London and Portland: Frank Cass.

—— 2003. The Civil–Military 'Gap' in Britain, *Journal of Strategic Studies*, 26:2, pp. 43–63.

—— 2006 (ed.). *Big Wars and Small Wars.* Abingdon: Routledge.

—— 2007. *Carl von Clausewitz's On War: A Biography.* London: Atlantic Books.

Strawson, John 1991. *Beggars in Red: The British Army 1789–1889.* London: Hutchinson.

Synott, Sir Hilary, KCMG 2009. Evidence to Chilcot Inquiry, 9 December.

Sztompka, Piotr 1999. *Trust: A Sociological Theory.* Cambridge: Cambridge University Press.

Tebbit, Sir Kevin, KCB CMG 2010. Evidence to Chilcot Inquiry, 3 February.

Thompson, E.P. 1991. *The Making of the English Working Class.* Harmondsworth: Penguin (4th edition).

Thorp, Arabella and Woodhouse, John 2009. *Immigration: Settlement and British Citizenship for Discharged Gurkhas and Commonwealth members of the Armed Forces*, House of Commons Library Standard Note SN/HA/4399.

Tipping Christianne 2008. Understanding the Military Covenant, *RUSI Journal*, 153:3, June, pp. 12–15.

Tootal, Stuart 2009. *Danger Close*. London: John Murray.

—— 2011. Interview with author, London.

Torpy, Air Chief Marshall Sir Glenn, GCB, CBE, DSO 2011. Evidence to Chilcot Inquiry, 18 January.

Torrance, Iain 1998. Ethics and the Military Community, *Strategic and Combat Studies Institute*, no. 34, July.

—— 2012. Correspondence with author.

Torrance, James B. 1970. Covenant or Contract? *Scottish Journal of Theology*, 23, pp. 51–76.

—— 1981. The Covenant Concept in Scottish Theology and Politics and its Legacy, *Scottish Journal of Theology*, 34, pp. 225–243.

Towle, Philip 2009. *Going to War: British Debates from Wilberforce to Blair*. Basingstoke: Palgrave Macmillan.

Tresch, Tibor Szvircsev and Leupracht, Christian (eds) 2010. *Europe Without Soldiers?* Kingston Ontario: McGill-Queens University Press.

Ulster Covenant 1912. Available at: www.proni.gov.uk/index/search_the_archives/ulster_covenant.htm (accessed 4 April 2010).

University Hospital Birmingham NHS Foundation Trust 2007. *Annual Report and Accounts*, 2006–2007. Available at: http://www.uhb.nhs.uk/Downloads/pdf/AnnualReport06–07.pdf (accessed 22 October 2011).

Urban, Mark 2003. *Rifles: Six Years with Wellington's Sharpshooters*. London: Faber.

US Chamber of Commerce 2011. *Defense Trade: Keeping the United States Safe and Competitive*, 2007 (Mathew Kazmierczak and Michaela Platzer). Available at: http://www.uschamber.com/sites/default/files/issues/defense/files/defensetrade.pdf (accessed 13 November 2011).

van Creveld, Martin 1991. *The Transformation of War*. New York: Free Press.

Vasquez III, Joseph Paul 2011. More than Meets the Eye: Domestic Politics and the End of British Conscription, *Armed Forces and Society*, 37:4, pp. 636–656.

Ven, Cyril van der 2007. The Psychological Contract: It is a Big Deal, in *Recruiting and Retention of Military Personnel*, NATO. Available at: www.rta.nato.int/pubs/rdp.asp?RDP=RTO-TR-HFM-107 (accessed 5 May 2010).

von Clausewitz, Carl 2007. *On War* (trans. Michael Howard and Peter Paret (eds); Introduction Beatrice Heuser). Oxford: Oxford University Press.

Wahrman, Dror 2001. The English Problem of Identity in the American Revolution, *American Historical Review*, 106:4, October, pp. 1236–1262.

Waldman, Thomas 2009. *War, Clausewitz and the Trinity*, unpublished PhD thesis. Available at: wrap.warwick.ac.uk/2048/1/WRAP_THESIS_Waldman_2009.pdf (accessed 10 October 2010).

Walker, Rt Hon General Lord (Michael) of Altringham, GCB, CMG, CBE DL 2005. *Memo from APO/Chief of the Defence Staff to APS/Defence Secretary on Iraq /Afghanistan commitments, 19 September 2005*. Available at: http://www.iraqinquiry.org.uk/media/44491/190905pso-aps.pdf (accessed 13 November 2011).

—— 2010. Evidence to Chilcot Inquiry, 1 February.

Waltz, Kenneth N. 1954. *Man, The State and War*. New York: Columbia University Press (2001 edition).

Walzer, Michael 1977. *Just and Unjust Wars: A Moral Argument With Historical Illustrations*. New York: Basic Books (4th edition).

War Office 1822a. Return of the Names of the Officers of the Army who receive Pensions for the Loss of Limbs or for Wounds, 18 April. National Army Museum Collection, London SW3, 8211/41/13.

—— 1822b. Return of the Names of Officers of the Army and of the Foreign Corps late in His Majesty's Service Who Receive Pensions for the Loss of Limbs or for Wounds, 18 April, National Army Museum Collection, London SW3, 8211/41/43.

Ware, Vron 2010. Lives on the Line, *Soundings*, 45, Summer, pp. 147–158 (republished in *Eurozine*, 28 Sept 2010).

Waters, Christopher P.M. 2008. Is the Military Legally Encircled? *Defence Studies*, 8:1, March, pp. 26–48.

Watt, Nicholas, Carvel, John and White, Michael 2008. Give Armed Forces Priority Access to Health and Education Services, Says Defence Chief, *The Guardian*, 25 June. Available at: http://www.guardian.co.uk/uk/2008/jun/25/military.defence (accessed 12 September 2010).

Westminster, Major General Gerald Grosvenor, The Duke of, KG CB OBE TD CD DL 2012. Interview with author, Warminster, Wiltshire.

Wheatley, Jane 2007. Growing Army of Soldiers' Wives Go to War over their Squalid Homes, *The Times*, 5 January, p. 30.

Wheeler, Nicholas 2000. *Saving Strangers: Humanitarian Intervention in International Society.* Oxford: Oxford University Press.

Wheeler, General Sir Roger, GCB CBE 2011. Interview with author, London.

White, Michael 2007. 'Devil in Detail' of Home Loan Aid to Military Personnel Say Tories, *The Guardian*, 29 December. Available at: http://www.guardian.co.uk/uk/2007/dec/29/politics.military (accessed 15 July 2010).

Wiltshire Gazette and Herald, Army Chief Pays Tribute to Town's Honour (online edition), 2 November 2007. Available at: http://www.gazetteandherald.co.uk/archive/2007/11/02/Poppy+Pride+News+%28swindon_news_poppy%29/1803954.Army_chief_pays_tribute_to_town_s_honour/ (accessed 15 May 2010).

Wither, James K. 2009. Basra's not Belfast: the British Army, 'Small Wars' and Iraq, *Small Wars & Insurgencies*, 20:3–4, pp. 611–635.

Wolseley, Lt General Sir Garnet 1882. *The Soldier's Pocket Book for Field Service.* London: Macmillan.

Worden, Blair 2009. *The English Civil Wars.* London: Weidenfeld & Nicolson.

YouGov/Daily Telegraph 2006. *Britain's Armed Forces in Iraq and Afghanistan.* October. Available at: http://today.yougov.co.uk/sites/today.yougov.co.uk/files/YG-Archives-Ira-dTel-BritainASIraqAfghan-061030.pdf (accessed 15 January 2011).

Index

For Product Safety Concerns and Information please contact our EU
representative GPSR@taylorandfrancis.com
Taylor & Francis Verlag GmbH, Kaufingerstraße 24, 80331 München, Germany

www.ingramcontent.com/pod-product-compliance
Ingram Content Group UK Ltd.
Pitfield, Milton Keynes, MK11 3LW, UK
UKHW011003180926
475701UK00004B/541
* 9 7 8 1 1 3 8 2 7 2 3 0 9 *